SpringerBriefs in History of Science and Technology

The *SpringerBriefs in the History of Science and Technology* series addresses, in the broadest sense, the history of man's empirical and theoretical understanding of Nature and Technology, and the processes and people involved in acquiring this understanding. The series provides a forum for shorter works that escape the traditional book model. SpringerBriefs are typically between 50 and 125 pages in length (max. ca. 50.000 words); between the limit of a journal review article and a conventional book.

Authored by science and technology historians and scientists across physics, chemistry, biology, medicine, mathematics, astronomy, technology and related disciplines, the volumes will comprise:

1. Accounts of the development of scientific ideas at any pertinent stage in history: from the earliest observations of Babylonian Astronomers, through the abstract and practical advances of Classical Antiquity, the scientific revolution of the Age of Reason, to the fast-moving progress seen in modern R&D;
2. Biographies, full or partial, of key thinkers and science and technology pioneers;
3. Historical documents such as letters, manuscripts, or reports, together with annotation and analysis;
4. Works addressing social aspects of science and technology history (the role of institutes and societies, the interaction of science and politics, historical and political epistemology);
5. Works in the emerging field of computational history.

The series is aimed at a wide audience of academic scientists and historians, but many of the volumes will also appeal to general readers interested in the evolution of scientific ideas, in the relation between science and technology, and in the role technology shaped our world.

***

All proposals will be considered.

Alessandro De Angelis

Galileo and Satellite Navigation

 Springer

Alessandro De Angelis
Delegation of Italy to the OECD and University of Padua
Paris, France

ISSN 2211-4564   ISSN 2211-4572   (electronic)
SpringerBriefs in History of Science and Technology
ISBN 978-3-031-78798-0   ISBN 978-3-031-78799-7   (eBook)
https://doi.org/10.1007/978-3-031-78799-7

Translation from the Italian language edition: "Galileo e la navigazione satellitare" by Alessandro De Angelis, © The Author(s), under exclusive license to Springer Nature Switzerland AG 2024. Published by Castelvecchi. All Rights Reserved.

This Springer imprint is published by the registered company Springer Nature Switzerland AG
The registered company address is: Gewerbestrasse 11, 6330 Cham, Switzerland

If disposing of this product, please recycle the paper.

Foreword

Until the end of the fifteenth century, sailors generally navigated by keeping close to the mainland. With the Portuguese explorations of the oceans, this was no longer possible. For a time, Iberian ships followed the coast of Africa, exploring its contours. But starting with the expeditions of Christopher Columbus in 1492 and Vasco da Gama in 1498, sailors were navigating for weeks on the open sea without seeing land. How did they know where they were and whether they were on the right course? Latitude (i.e., the angular distance from the equator) is easy to measure by comparing the ship's position with the North Star. But what about longitude, that is, the angular distance from an arbitrary meridian (today, the Greenwich meridian passing through London)? How could this be determined? For nations engaged in trade with the Indies, finding longitude at sea was crucial: only knowing both latitude and longitude allowed for accurate geolocation.

At the end of the sixteenth century, the Spanish Crown established a large prize in hopes that someone would find a solution. This initiative was followed by similar actions from the Dutch, French, and English governments. Shortly after the discovery of Jupiter's moons, Galileo Galilei realized that they provided a clock visible from any observation point. When a moon enters the shadow cone of the planet, it disappears very quickly. If a navigator on the open sea could note the local time of an eclipse and compare it with the local time when it was expected to occur at a reference position in Europe, the difference in times, and thus the longitude, could be easily measured.

In 1612, Galileo entered into negotiations with Spain, offering Spanish navigators an instrument to calculate the times of eclipses of the moons and telescopes with which to make the observations. However, the Spaniards discovered a serious problem with his method. To observe the moons, a relatively powerful telescope was necessary; given the narrow field of view of Galileo's telescope, it was impossible to make the observation from the deck of a ship on the open sea. Galileo made some attempts by fixing a telescope to a helmet, but this approach only worked with telescopes of rather low power. The Spaniards were not impressed by Galileo's method, and the negotiations ultimately failed. Galileo took up the problem again after his condemnation in 1633, and this time negotiated with the Dutch States

General. The Dutch government was unable to reach a conclusion before Galileo's death.

This little-known story, fascinating in many ways both from a scientific perspective and for its dramatic nature, is recounted here by Alessandro De Angelis, a physicist who has already successfully tackled the history of science and Galileo's contributions. The book explains the problem by placing it in a historical context, and for the first time offers an exhaustive review of Galilean documents on the subject. Galileo dedicated more than twenty years of his life to research on geolocation. When Galileo died, his student Viviani predicted that this method of localization, using the moons of Jupiter or others, would one day be followed by all of humanity. And indeed, today we are accustomed to using our cell phones to determine our geographical position by exploiting global satellite navigation systems. There are four in the sky: one American, one Russian, one Chinese, and one European. Each of these positioning systems relies on about thirty artificial satellites, and the European one is named after Galileo.

This book shows us how the imagination of a scientist like Galileo was able to go beyond the technical possibilities of his time and imagine an invention that would only become feasible three hundred and fifty years after his death. It also shows us how history and science are closely linked. And above all, it entertains and educates us, taking us effortlessly on a journey that spans from the first ideas over four centuries ago to today's technologies. The research into sources is exhaustive, with an extensive appendix dedicated to original sources.

The book reads like a novel and will be appreciated by anyone interested in how we have come to navigate safely and easily not only on the seas and land but also in the sky.

University of Padua   William R. Shea
Padua, Italy

Preface

Soon after discovering Jupiter's satellites, Galileo Galilei realized that their mutual positions provided an absolute clock that could be consulted from Earth, and the possible uses of this clock became an obsession that haunted him until he died. A navigator on the high seas would have been able to note the local time of an eclipse of one of Jupiter's satellites (on average, there are two or three each night) and compare it with the time it was expected to occur at the European reference point; the difference would provide the ship's longitude, which, combined with a well-known technique for calculating latitude, would allow geolocation. In 1612, Florentine diplomacy began negotiations with the Spanish Crown to provide Iberian navigators with tables (from which an instrument was also derived) of Galilean invention to calculate eclipses of Jupiter's satellites and telescopes with which to make observations. It took the Spaniards six years to evaluate the idea, concluding that it was impossible to perform the observation from the deck of a ship on the high seas because of roll and pitch. Galileo took up the problem again after his conviction (which occurred in 1633), this time negotiating through intermediaries with the States General of the Netherlands. While Galileo was rewarded (pending the final evaluation of the project) with a valuable gold necklace, which the Inquisition forbade him to accept, the Dutch technical committee had doubts in line with the Spanish committee.

Satellite navigation, technically impossible at the time of Galileo because of Jupiter's satellites' considerable distance and relative angular proximity and the difficulties of pointing telescopes, has become feasible in the space age. Of the four constellations of geolocating satellites in orbit today, the European constellation is named after Galileo himself.

This book tells the story of the Galilean proposal for geolocation, framing it historically and astronomically between the "before" (the techniques attributed to the Greeks and perfected by medieval astronomers) and the "after" (the arrival on the scene of precision clocks; then, in the twentieth century, the wireless transmission of electromagnetic signals; and finally, space technology).

The book is based primarily on the twenty volumes of the beautiful National Edition of the Works of Galileo, published between 1890 and 1909 by Antonio

Favaro. Keeping with a widespread tradition among Galileo scholars, I will refer hereafter to this book simply as the *Works*. I also make use of the fourth volume of the updated edition of the *Works*, edited by Michele Camerota together with Patrizia Ruffo; this volume, published in 2019, presents some unpublished documents on the contacts between Galileo and the Spanish government for the sale of the Galilean method of determining longitude at sea.

The documents cited, Galileo's letters, and those otherwise relevant in the context of the book are mostly taken from the *Works*. I always make a clear reference to the original edition, giving in square brackets, where possible, the progressive letter number in the *Works*, preceded by the volume number (e.g., [11/757]), otherwise the volume and page (e.g., [A4/p79], where A4 indicates the fourth volume of the Update). I have translated the texts from Italian, Castilian, Latin, and French, but always with a clear reference to the original. In short, the reader who is particularly curious or interested in philology will easily find the necessary references to satisfy his curiosity. I have also relied, to a lesser extent, on some of the texts listed in the bibliography.

The occasion for writing this book was the organization by the Italian Permanent Delegation to the International Organizations in Paris of the 2024 "Italian Research Day in the World": the subject was geolocation. The event, organized in collaboration with the Museo Galileo in Firenze and Sorbonne Université Paris, comprised the exhibition "Galileo and satellite navigation" held at the "La Passerelle" gallery in the Pierre et Marie Curie Campus from June 13 to June 28, 2024. The exhibition was replicated in the Italian Institute of Culture in Prague from October 21 to 31, in the Italian Institute of Culture in Amsterdam from December 9 to 13, and is going to be replicated again in the Italian Institute of Culture in Stockholm, and at the Perimeter Institute in Waterloo (Canada).

Paris, France   Alessandro De Angelis
August 2024

Contents

Chapter 1
Determine Where We Are

Earth is the third of the eight planets in the solar system, in order of distance from the Sun. Mercury and Venus are inner planets, while Mars, Jupiter, Saturn, Uranus, and Neptune are outer planets. Our planet orbits the Sun yearly at an average distance of about 150 million kilometers; this distance is called an astronomical unit. Light from the Sun takes about 500 s (just over 8 min) to reach us.

The Earth is roughly a sphere with a radius of 6400 km. It rotates daily on itself around an axis that intersects the Earth's surface at two points called the poles: North and South. The North Pole is such that an observer sees the Earth's rotation counterclockwise with his feet to the south and his head to the north. The equator is the largest circle around the Earth, drawn perpendicular to the axis of rotation, and divides the Earth into two hemispheres, northern and southern.

1.1 The Coordinates on Earth: Latitude and Longitude

Like any point on a surface, a point on the Earth's surface can be defined by two coordinates. Latitude and longitude, both measured in degrees, are the most used coordinates. They use imaginary lines called meridians and parallels that geographers use to divide the space of the Earth's surface (Fig. 1.1).

Latitude (from Latin *latitudo*, width) is the geographic coordinate that indicates how far north or south a point on the Earth's surface is from the equator. It is an angle from minus 90 degrees (also called 90 degrees south, or 90 degrees S, or 90° S) at the South Pole to 90 degrees (also called 90 degrees north, or 90 degrees N, or 90° N) at the North Pole, with 0 degrees at the equator. Lines of constant latitude, or parallels, run east-west as circumferences parallel to the equator.

Longitude (from Latin *longitudo*, length) indicates the east-west position of a point on the Earth's surface. It is an angular measurement, usually expressed in degrees. Meridians are imaginary semicircular lines that run from pole to pole, connecting points of the same longitude.

© The Author(s), under exclusive license to Springer Nature Switzerland AG 2024
A. De Angelis, *Galileo and Satellite Navigation*, SpringerBriefs in History of Science and Technology, https://doi.org/10.1007/978-3-031-78799-7_1

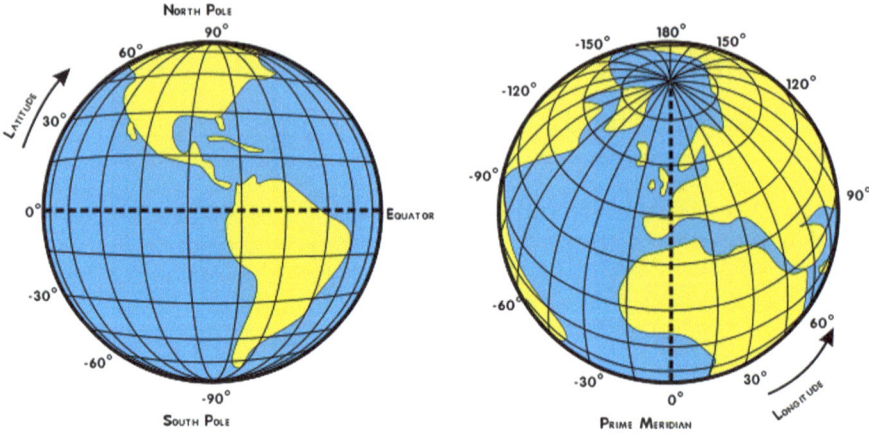

Fig. 1.1 Definition of latitude and longitude for a point on Earth. From Wikimedia Commons

Unlike latitude, which has an objective starting point because the Earth rotates on itself, the meridian of zero longitude is arbitrary. Beginning in the fifteenth century, European maritime powers chose a zero meridian that was convenient for them to write logbooks and nautical charts: for example, that of the island of Terceira in the Azores for Portugal, a meridian running through the Canary Islands, or that of Toledo, for Spain. The King Louis XIII of France convened a conference of cartographers and astronomers from all over Europe in April 1634 to establish a single reference. Sanctifying a tradition that dated back to Claudius Ptolemy, a Greek astronomer of the second century A.D., they chose to adopt the meridian of the Isle of Hierro (in the Canary Islands), which for Ptolemy was the westernmost point of the inhabited lands. Immediately, Flemish cartographers, the best of the time, followed this convention so that most historical maps still refer to it. In 1718, the Franco-Italian astronomer Cassini drew the meridian of Paris, and the first topographic map of France was drawn in reference to it. The French attempted to establish this meridian as the universal meridian, but their attempt was unsuccessful. Even today, walking through the streets of Paris, one can see 135 plaques indicating the local meridian. In 1738, the British Empire adopted the Greenwich meridian, which passes near the Royal Observatory in southeast London, as its zero meridian. As you can imagine, there was a great deal of confusion until 1884, when an international conference in Washington adopted the Greenwich meridian as the international zero meridian. This reference prevailed because of its association with the Royal Observatory and the importance of the British Empire in navigation and cartography during the eighteenth and nineteenth centuries. Positive longitudes are those east of the zero meridian up to 180 degrees (also called 180 degrees east, or 180 degrees E, or 180° E), and negative longitudes to the west to minus 180 degrees (also called 180 degrees west, or 180 degrees W, or 180° W).

Starting from Rome (which has a latitude of about 42 degrees north and a longitude of about 12.5 degrees east), one degree of longitude corresponds to about

85 km in the east-west direction, and one degree of latitude corresponds to approximately 110 km in the north-south direction.

Because of the Earth's rotation, there is a close relationship between longitude and time. Local time varies with longitude: a 15-degree difference in longitude corresponds to a 1-h difference in local time due to the difference in position relative to the Sun (1 day, or 24 h, corresponds to 360 degrees). Since the difference in longitude between Rome and Paris is about 10 degrees, astronomical time in Rome (say, roughly, the hour at which the Sun rises) differs by about 40 min from that of Paris.

Because Italy and France have decided to use the same time, called Central European Time or CET, the Sun rises on average 40 min later in Paris than in Rome (and of course sets 40 min later). The charming city of Santiago de Compostela, in Galicia, is one of the westernmost cities in Central European Time. It is 21 degrees longitude from Rome, which means that, on average, the Sun rises (and sets) almost an hour and a half later there. Keep that in mind when you program your evenings!

Measuring latitude, i.e., the height relative to the equator, is relatively simple and can be done accurately in a short time without the need for ground reference points (and thus even from a ship in the middle of the ocean). In the northern hemisphere, the axis of rotation projects approximately on the polar star (Fig. 1.2).

With a simple instrument such as a quadrant (Fig. 1.3), we can measure the angle between the pole star and the horizon; this, based on simple properties of triangles (Fig. 1.2), is the measure of latitude—then we make minor corrections related to the deviation between the pole star and the North Pole. A quadrant (like its successor, the sextant) is a navigational instrument used to measure the angle between celestial objects, such as between celestial objects, such as the Sun, Moon, or stars, and the horizon. It consists of a graduated arc, usually calibrated in degrees, and a sighting

Fig. 1.2 Measurement of latitude. A sailor at position P measures the angle (ϕ) between the horizon and the direction of the North Star. Based on simple geometric considerations it can be demonstrated (see figure) that this is equal to the angle between the position and the equator, i.e., the latitude

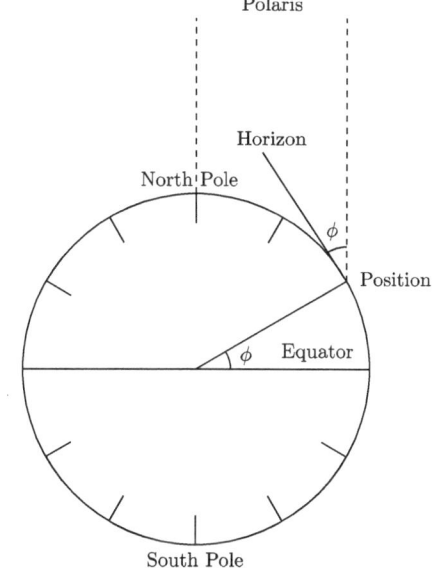

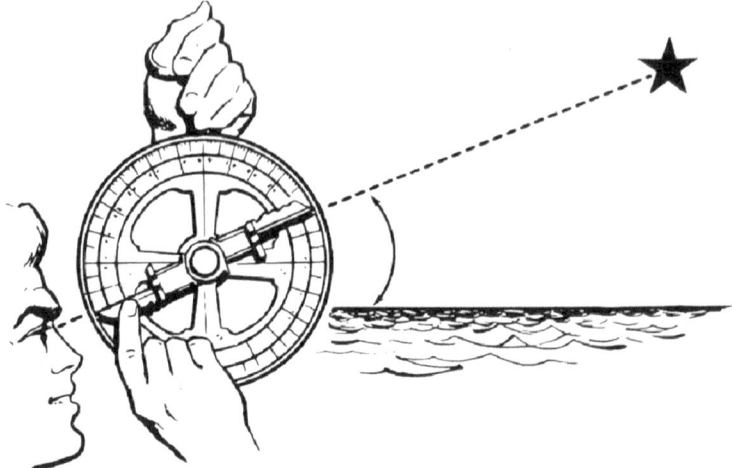

Fig. 1.3 Simplified diagram of the operation of a quadrant (or of a sextant, or of an octant). A device allows to maintain a fixed axis with respect to the horizon

mechanism that allows it to be aligned with the horizon or another reference direction (Fig. 1.3). An instrument with a similar function but a more complex design, the astrolabe, was used by Arab astronomers (Fig. 1.4). It consists of a fixed graduated circle (mother) on which a projection of the celestial sphere is reproduced, and a second circle (net), movable and perforated, sliding on the first, which reproduces the positions of the main reference stars.

Quadrants and sextants have been used in maritime navigation for centuries. The use of quadrants has been documented in the literature of various ancient civilizations, including those of Greece, the Roman Empire, and the Arab culture (which, as mentioned above, also made great use of the astrolabe). However, the earliest archaeological evidence of quadrant-like instruments dates to ancient Egypt and Mesopotamia, before 1500 BC. These instruments were later perfected in the age of the great navigators beginning in the fifteenth century AD.

Of course, even in the Southern Hemisphere, where the Pole Star cannot be seen, we can use the position of the stars near the Southern Cross to find a point that indicates the South Pole, and from that point, we can use the same method as above to find the Southern Longitude.

Determining longitude, on the other hand, is more complicated (as we can understand from the map in Fig. 1.6, which shows great accuracy in latitude determination and only a very approximate knowledge of longitude). Several methods have been proposed. We will look at three of them now, and two more will be discussed in the next section.

– Clock method. By comparing local time with the time of the reference meridian, we can determine longitude (Fig. 1.5). As we have seen, the Sun rises 40 minutes later in Paris than in Rome, so we can say that the longitude of Rome is 10 degrees east of that of Paris. This measurement requires that two clocks in Rome

Fig. 1.4 An astrolabe. From Wikimedia Commons

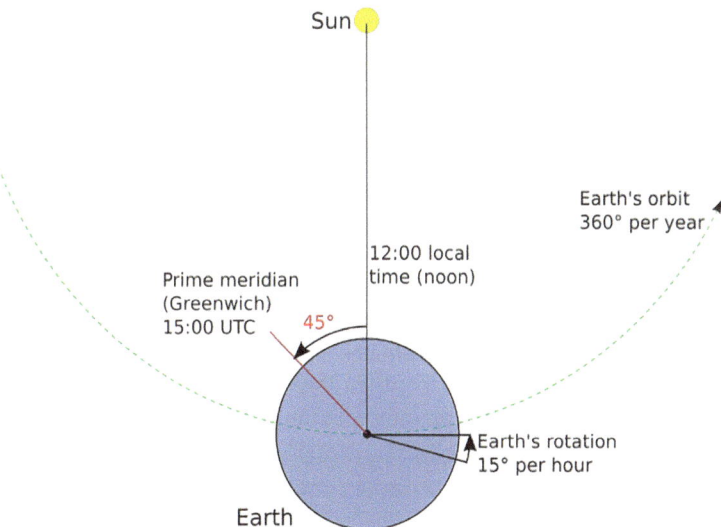

Fig. 1.5 Longitude relative to a reference meridian (such as the Greenwich meridian) can be calculated by comparing the time given by the Sun's position with the reference time

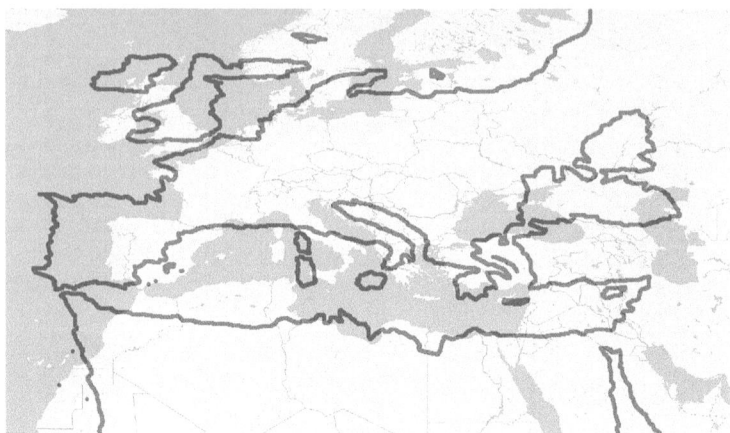

Fig. 1.6 Ptolemy's map of the Mediterranean superimposed on a modern map (from Wikimedia Commons)

and Paris are synchronized and that the time of sunrise on a given day in each place can be calculated well. Instead of dawn, one can also use the time of local (astronomical) noon, which is the time when the Sun is highest in the sky. This is difficult to determine directly since the Sun's apparent motion is nearly horizontal at noon, but you can take the midpoint between two times when the Sun was at the same height above the horizon.

– Lunar eclipse method. Alternatively, an estimate of the difference in longitude can be obtained from a celestial event that is visible from two locations at the same time (and thus, of course, at different local times), such as, to a good approximation, a lunar eclipse, which is seen almost simultaneously over the entire half of the Earth that can observe it. The principle is simple. Suppose I know the time of an eclipse in Lisbon and am on a ship in the middle of the ocean. In that case, I can compare my local astronomical time (which I determined the day before from the Sun's position) with the time in Lisbon. But eclipses are rare astronomical events.

– Method of comparison with nearby places. With a bit of patience, we can determine the longitude of a fixed location on land with respect to the zero meridian at Greenwich, for example, by waiting for a lunar eclipse. We can then mark it on a map. Before the age of satellites, people relied on compasses and distance measurements to determine their position on land by referring to the coordinates of points on maps whose coordinates had previously been well measured. Triangulation measures angles between three visible landmarks whose coordinates are known. From these, an interpolation is performed. We will discuss this technique in more detail later.

1.2   First Solutions to the Problem of Localization

As mentioned above, the problem of determining latitude is relatively simple, even in the absence of nearby landmarks with specific coordinates, and can be solved quickly, even at sea or in the air.

Longitude is a different story. Measuring the longitude of a point takes a long time, and there is not an easy solution for moving systems such as ships or airplanes.

The fifteenth century was the dawn of the age of the great explorers. Brave men traveled the seas, rivers, and unknown lands, opening new trade routes and enriching our knowledge of the world. The Genoese Christopher Columbus (1451–1506) sailed westward across the Atlantic Ocean in search of a new route to Asia but instead discovered the American continent in 1492. The Portuguese Vasco da Gama (1469–1524) was the first European to sail directly to India, rounding the Cape of Good Hope south of Africa in 1498. The Portuguese Fernando Magalhaes (1480–1521) led the first circumnavigation of the globe from 1519 to 1522. Although he died during the voyage, he proved that the Earth is a sphere and not a flat disk.

Incidentally, the 18 Magalhaes expedition survivors, who returned to Spain after sailing around the world, were astonished to find that the log kept by the highly accurate Antonio Pigafetta marked 1 day less than the Andalusian calendar. The calendars were out of sync during the voyage—a dramatic example of the relationship between longitude and time. Aboard ship, time was measured by reference to local (astronomical) time, which meant a loss of 1 day in circumnavigating the planet.

It was beneficial on these dangerous voyages to know where you were.

The technique of comparing local time with the time of the reference meridian required that those who had to measure their longitude had a clock synchronized with the reference meridian. At the end of the sixteenth century, technology was not advanced enough, and there were no portable clocks accurate enough to allow measurement except with substantial errors. Moreover, once the clock "lost" time, one also lost the ability to orient oneself. Recall that a 1-h mistake in the reference clock, easy to have after a month of navigation, results in an error of 15 degrees in longitude.

In addition to the three techniques described in the previous section (comparison of local time with that of the reference meridian, lunar eclipses, and comparison with the positions of visible points), two other techniques were used.

- Magnetic field method. One technique made use of the geomagnetic field. The magnetic north pole (measured with a compass) differs from the geographic north pole (measured, for example, with the pole star, with minor corrections); the difference between the two depends on longitude. This method is also quite imprecise, with typical errors of about fifteen degrees.
- Lunar distance method. The lunar distance method, based on measuring the angular distance between the Moon and a star or other known celestial body, was the most widely used. This distance was then compared to the motion of the Moon to obtain a measure of longitude with respect to a meridian reference axis.

The proposal dates to Ptolemy. Amerigo Vespucci (1454–1512), a Florentine who served the kingdom of Portugal and then the kingdom of Castile, was one of the first to use this method in his navigation along the coast of Venezuela in 1499. Johannes Werner, astronomer and pastor of St. John's Church in Nuremberg, one of the founders of modern trigonometry, studied and theorized the method of lunar distances and perfected it by publishing his results in the form of a lunar almanac in 1514. He proposed to map the position of the stars along the lunar orbit and compare the local time at which the eclipse of the stars was sighted at the observer's meridian with the predicted time at the Greenwich meridian. In this way, the difference in longitude could be calculated. At that time, however, the positions of the stars were not well known, and the laws of motion of the Moon had yet to be discovered, making it difficult to predict its orbit accurately. In addition, sailors had to make non-trivial calculations, such as correcting for the Moon's orbital motion. As a result, the method of determining lunar distances was not very accurate (about fifteen degrees, or 1200 km of uncertainty at the latitude of Rome) until further advances were made in understanding astronomy and collecting more accurate celestial data.

Chapter 2
Galileo's Solution: The Satellites of Jupiter

In 1610, nearly a century after the publication of Werner's Lunar Almanac, Galileo Galilei discovered in his garden in Padua what he thought might be the coveted ideal celestial clock.

Galileo, the first person to point a telescope at the sky, discovered the Moon's mountains, sunspots, the phases of Venus, a ring around Saturn (which he mistook for two moons near the planet), and a constellation of four satellites orbiting the planet Jupiter, moving in the same way that planets move around the Sun. These four satellites are now called the Galilean or Medicean Moons (Galileo dedicated the discovery to his Florentine patron, Grand Duke Cosimo II de' Medici).

The first observation was made on January 7, 1610. Galileo observed three small points near Jupiter. He continued to observe Jupiter for about a month (weather permitting), noting in his notebook the position of the dots, which turned out to be satellites orbiting the planet. At one point, these dots became four (in the early days of observation, one of them was eclipsed by Jupiter itself). It is still exciting to consult Galileo's notebook of observations (Fig. 2.1).

Beginning in April 1611, Galileo, who had moved to Florence, measured the periods of revolution of the Medici satellites around their planets. The idea of using these planets' very frequent transits and eclipses to solve the problem of determining longitude at sea came to his mind almost immediately.

Let us pause for a moment and summarize the properties of Jupiter and its satellites.

2.1 Jupiter and Its Moons

Jupiter is the fifth planet in our solar system, counting from the Sun outward. It is located at an average distance of about 778 million kilometers from the Sun, or about five times the average distance between the Earth and the Sun. As one of the outer planets, its orbital period is longer than a solar year, as predicted by Kepler's

A. De Angelis, *Galileo and Satellite Navigation*, SpringerBriefs in History of Science and Technology, https://doi.org/10.1007/978-3-031-78799-7_2

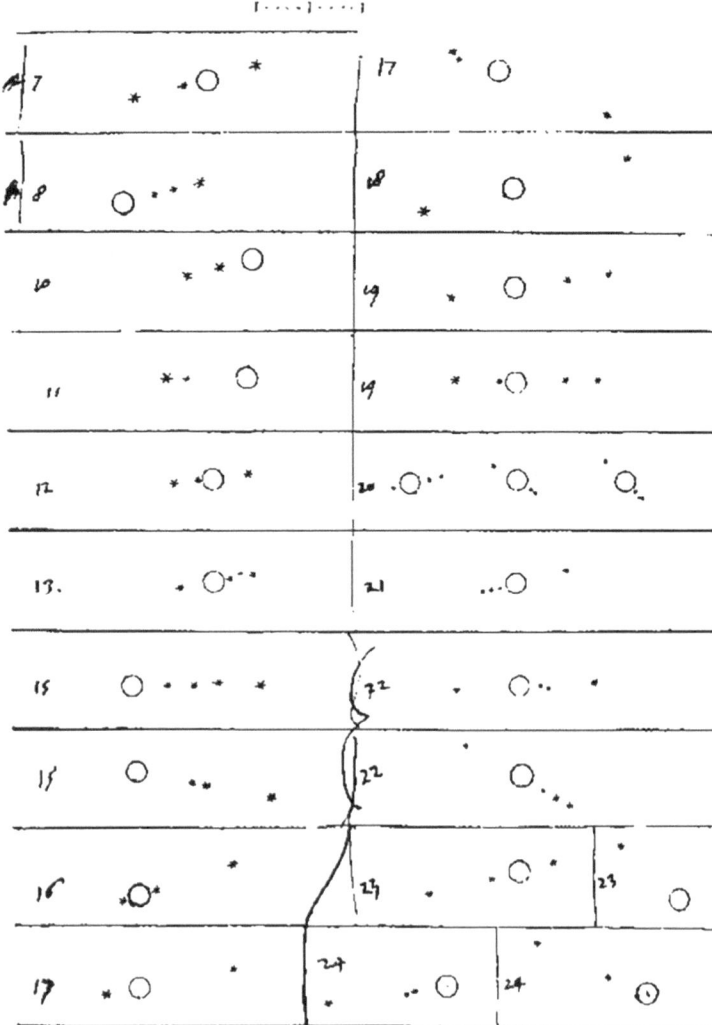

Fig. 2.1 Observations of Jupiter's moons from Galileo's notes taken between January 7 and January 24, 2010 (Biblioteca Nazionale Centrale, Florence, page 70r of the manuscript Gal. 50. Credits: Ministero della Cultura, Italy)

third law relating orbital periods to distances from the Sun: the duration of a "Jovian year" is about 11 Earth years. Jupiter is the largest planet in the Solar System with a radius of about 72,000 km, about 11 times that of the Earth and one-tenth that of the Sun. Its mass is more than three hundred times that of the Earth and one-thousandth that of the Sun. It is clearly visible almost every night and is the brightest object in the night sky after the Moon and Venus. Because of its size, it has played a prominent role in the religious beliefs of many cultures, including the Babylonians,

Greeks, and Romans, who identified it with the ruler of the gods. His astrological symbol, which later became astronomical ($2\!\!\!\downarrow$), is a stylized representation of lightning, the main attribute of this deity.

Jupiter can be considered a "failed star" because of its size and Sun-like composition—it is composed mainly of hydrogen and helium, like the Sun. It does not have a defined solid surface like "inner" rocky planets like Earth. If Jupiter had had the opportunity to increase its mass by a factor of 100, its core would have reached the temperature and pressure conditions necessary to initiate the chain fusion reactions of hydrogen into heavier elements. Nuclear fusion would have made it a star, and the solar system would have become a binary star system, like half the star systems we know of in the universe.

Jupiter's atmosphere is characterized by a series of cloud bands surrounding the planet in an east-west direction. The bright bands are called zones, while the dark bands are called bands. These atmospheric phenomena are caused by Jupiter's rapid rotation (it rotates on itself about every ten hours) and atmospheric currents. A famous feature of Jupiter is the Great Red Spot, a giant oval storm large enough to contain two or three planets like Earth, one of the most persistent and visible features.

Jupiter has been the subject of several space missions. The Voyager 1 spacecraft made a close flyby of Jupiter in 1979, returning detailed images and data about its atmosphere and moons. The Galileo spacecraft, launched in 1989, orbited Jupiter for nearly eight years, studying its atmosphere, magnetosphere, and Galilean moons. Other missions, such as the Juno spacecraft launched in 2011, continue to improve our understanding of the planet.

Jupiter's intense gravitational field—remember that gravitational attraction is proportional to the mass of an object—affects the structure of the solar system, disrupting the orbits of the other planets and "cleaning" some of the debris that might hit the innermost planets by attracting such debris to the planet. Because of its strong gravitational pull, Jupiter has an extensive system of moons.

As of February 23, 2023, 95 moons have been observed orbiting Jupiter.

The Galilean or Medicean moons are the four most massive moons. Galileo Galilei, who discovered them in Padua in January 1610 during one of the first observations with the telescope he perfected, dedicated his discovery to the Grand Dukes of Florence. Today, the moons are named, in order of distance, Io, Europa, Ganymede, and Callisto, after the names of four lovers of Jupiter-three female and one male (Ganymede). The Medicean satellites have an apparent brightness near the limit of naked-eye visibility, which would theoretically make them visible to the naked eye (though probably not separable) were it not for the intense brightness of the planet towering above them. Indeed, the first record of the existence of a satellite structure around Jupiter is in a note by the Chinese astronomer Gan De, dated around 364 B.C., about a "reddish star" near Jupiter, perhaps observed by shielding the planet's light. However, the first definite observations of Jupiter's satellites were made by Galileo, who published his results in *Sidereus Nuncius* in March 1610. A historical rival of Galilei, the German Simon Marius (Simon Mayr), formerly a lecturer in Padua, independently discovered Jupiter's moons shortly after Galileo, perhaps even a few days later, but did not publish his results until 1614. However, the names in use today are those assigned by Marius.

Dozens of much smaller Jovian moons have been identified since 1892; they have been given the names of lovers (or other sexual partners) or children of the Roman god Jupiter or his Greek counterpart Zeus.

The four Galilean moons are by far the largest and most massive objects in Jupiter's orbit, while the remaining known moons and other debris together account for only 0.003 percent of the total orbital mass. They are easily visible with a small amateur telescope (Fig. 2.2) or even with good binoculars.

The four Medicean satellites have a radius of more than 1500 km (remember that the Earth's radius is about 6400 km).

With a diameter of 5300 km, Ganymede is the largest of Jupiter's satellites and the largest object in the Solar System except for the Sun and the seven planets. Mercury, although more massive, has a smaller diameter. It consists mainly of ice and rock, with a thin layer of atmosphere. Ganymede has been the subject of many space missions, including NASA's Galileo probe.

Callisto is Jupiter's second-largest moon and has a very thin atmosphere composed mainly of carbon dioxide. Its surface is very old and cratered, with a lack of significant geological activity. Callisto was also studied by the Galileo mission.

Io is the closest moon to Jupiter and is one of the most volcanically active places in the solar system. Its surface is covered by numerous volcanoes and has a thin atmosphere composed mainly of sulfur. Io has been the subject of several space missions, including the Galileo, Voyager, and New Horizons spacecrafts.

Fig. 2.2 Jupiter's satellites seen with a small amateur telescope costing about a hundred euros, with a focal length of 360 mm and an aperture of 50 mm

Europa, the smallest of the Medicean moons, is considered one of the most promising places to search for extraterrestrial life in the Solar System. Its surface is mainly ice, under which an ocean of liquid water is thought to exist. It has a very thin atmosphere and has been the subject of study by the Galileo mission, as well as proposals for future exploration missions.

The characteristics of Jupiter's moons are summarized in Table 2.1.

The study of Jupiter's moons has provided much valuable information about the formation and evolution of planetary systems. These moons are considered important targets for future space missions because they may offer additional clues to the presence of life in the Solar System.

The space exploration of Jupiter's satellites, beyond the results already achieved by the Galileo mission, includes a key mission: JUICE (JUpiter ICy moons Explorer), an instrument launched by the European Space Agency in 2023, aims to explore Ganymede, Europa, and Callisto. It will reach its destination in 2031.

A notable characteristic of the orbits of Jupiter's satellites is that their periods last for days. The scenario changes every day, as we have seen from Galileo's notebook of early observations. Eclipses of Jupiter and "overshoots" are common phenomena. And because of Jupiter's great distance from the Earth, each of these events provides a time that is virtually simultaneous on the surface of the Earth.

We notice that the periods of Jupiter's first three satellites are almost precisely in the ratio 1:2:4. Is this an accident? No: it is a mechanism explained by the Frenchman Pierre-Simon de Laplace in the eighteenth century called orbital resonance. The periodic alignment of the moons creates periodic interactions in the mutual gravitational attraction between the moons. The system could become unstable, and one of the moons could escape, but fortunately, there are stable solutions. One solution is orbital resonance, like the resonance of the strings of a musical instrument, which, when plucked, produces complex transient vibrations that die out after a few moments, leaving only harmonic frequencies, i.e., with periods in simple proportions.

2.2 Galileo's Idea

Before Galileo, one of the greatest geniuses in the history of physics (one of the five most important, according to Hawking) and, together with John Kepler, the most famous astronomer of his time, all the efforts of scientists had been in vain. No progress had ever been made with respect to the method of lunar distances proposed

Table 2.1 Properties of Jupiter's satellites

Satellite	Diameter	Distance from Jupiter	Period
Io	3643 km	421,700 km	1.8 days
Europa	3122 km	671,034 km	3.6 days
Ganymede	5262 km	1,070,412 km	7.2 days
Callisto	4821 km	1,882,709 km	16.7 days

by the ancients, especially the Greek astronomers Hipparchus and Ptolemy, and later perfected by Werner.

The most accurate method used to date by the great cosmographers to determine longitude used lunar eclipses. This method, proposed by Hipparchus in the second century BC, has many difficulties. The first difficulty is the rarity of the phenomenon. There are at most two visible lunar eclipses per year, sometimes only one and sometimes none. In addition, it is tough to accurately observe an eclipse's beginning, middle, or end. When the Moon begins to enter the Earth's shadow, the shadow is so faint and blurred that the observer is confused about whether the Moon has already started to touch the Earth. Thus, one could make an error of more than a quarter of an hour (about four degrees of longitude).

Galileo Galilei was not an expert in navigation, but like all scientists of his time, he was aware of the longitude problem. The problem, much studied because of the significant consequences for navigation that depended on its solution, aroused the interest of many brilliant minds, also because of the rewards offered by various naval powers to those who solved it.

Galileo discovered the moons of Jupiter in Padua in 1610, a year of significant changes for him: he moved from Padua to Florence, a move he had been pursuing for a decade, and he had to deal with organizational problems related to the settlement of his family. He also left his partner and had to find someone to take care of their three children, whom he had brought with him but could not support alone. So it was only in April 1611 that he had time to patiently observe and systematically study the motions of Jupiter's moons, calculating the periods of their orbits and counting the number of times "the small bodies disappeared behind the giant's shadow".

Although it was not possible to see Jupiter's moons during the day because the Sun's light obscured the planet, or when the planet was absent from the sky, or when the sky was not clear, eclipses of Jupiter's moons are common phenomena because the periods of these stars are only a few days long. Their occultations could therefore be measured more easily and accurately than those of the Moon. Galileo realized that by observing the transits and occultations of Jupiter's satellites, it would be possible to reduce the error in longitude to only a minute of an hour, or a quarter of a degree, thus gaining a considerable advantage.

Galileo produced accurate tables to determine the positions of the observed moons as a function of the time of observation, which he completed in 1612. These tables are commonly referred to in astronomical jargon as ephemerides (from the Greek, tables containing calculated values, day by day, of variable astronomical quantities). In his tables, Galileo accurately predicted the motion of the satellites. This was no easy task since the observed positions of the satellites and the apparent radii of their orbits are significantly affected by the Earth's annual motion around the Sun. Fortunately, Galileo was a Copernican, so he was fearless in changing the reference system for his calculations to use the Sun as the reference system!

An instrument called the giovilabium, or jovilabe (Fig. 2.3), an analog calculator that calculated time from positions and was named for its resemblance to an astrolabe, was also later constructed from Galileo's tables of positions. Werner's method could be used any night, but it needed to be more accurate. In addition to all the shortcomings of the eclipse method, it suffered from uncertainty in predicting the Moon's motion. Galilei's approach seemed unbeatable on paper.

As we shall see, the devil that would not allow Galileo to assert his idea lays in the not insignificant detail of the difficulties of observation. How could one aim at the moons of Jupiter from the deck of a rolling, pecking ship with the crude telescopes of the time? Galileo did not realize the problem at first, but he soon set about

Fig. 2.3 A jovilabe from middle seventeenth century built on Galileo's project, on display in the Museo Galileo in Florence (inv. 3178)

solving it with the originality and genius that distinguished him in all his work. This adventure kept the Tuscan genius busy for almost thirty years and is the subject of the rest of this chapter.

2.3   Galileo Unsuccessfully Proposes His Idea to the King of Spain

In 1612, having determined the orbits of Jupiter's satellites, Galileo was convinced that he had solved the problem of locating a point on the Earth's surface, and he told Grand Duke Cosimo II, his friend and protector, so that he could compete for the generous Spanish prize. He convinced Cosimo both because Cosimo had been his student and had esteem, gratitude, and admiration for him, and because of a fortunate coincidence.

Galileo's proposal could not have come at a better time. In June 1612, the government of Madrid had asked Grand Duke Cosimo II to arm and sail certain galleons in the port of Livorno for protection against pirates. Tuscany could thus include Galileo's proposal in the negotiations.

The history of the Florentine fleet is fascinating. The last venture of Grand Duke Ferdinand, son of Cosimo I and Eleanor of Toledo and father of Cosimo II, who succeeded him, had ended in the 53rd year of his life. At an advanced age, fascinated by his mother's tales of the wealth of the Spanish colonies, he had decided that Tuscany should also become a colonial power, as much as and more than Spain, and that he would be its emperor. He had armed a fleet and organized an exploratory expedition to northern Brazil and the Guianas. The galleon Santa Lucia had left the port of Livorno in January 1608, commanded by the Englishman Thornton and loaded with convicts. The expedition's captain and the sailors returned in 1609, bringing much information and many gifts for the prince, predominantly plants, tropical birds, and some natives. But they found that the prince had been dead for several months. The colonial project was promptly canceled by Grand Duchess Christina, who had lovingly tolerated her husband's mattocks but was not enthusiastic about them. Christina acted as regent until Cosimo II reached the age of twenty-one, and she had eight other younger children to care for.

In exchange for the aid of the Tuscan fleet to Spain, Grand Duke Cosimo II demanded certain trade privileges with the Indies, including the ability of the Tuscan fleet to make expeditions to the Indies without interference.

Galileo's proposal was thus synergistic with the negotiations. The Grand Duke mentioned it to the Tuscan first secretary (prime minister), Belisario Vinta, who wrote to the Tuscan ambassador in Madrid, Orso Pannocchieschi, Count of Elci, on September 7, 1612, asking him to act as an intermediary (see Appendix A.1).

Through the ambassador, the Tuscan government offered the King of Spain Galileo's technique for measuring longitude and telescopes of superior quality to those of Iberian manufacture since they were based on unsurpassed Venetian optics.

The official communication from the Tuscan government to the ambassador in Madrid for transmission to the King of Spain, entitled Proposal of Longitude, was written by Galileo himself. However, the scientist naturally refers to himself in the third person (Appendix A.2). The original document, written in his hand, is preserved in the State Archives in Florence.

To persuade the Spanish government to consider favorably Galileo's proposal and that of the Grand Duchy's policy, the ambassador also suggested "that the method of measuring longitude be introduced and taught at every hour of the night and almost all the time of the year; that those who understand navigation affirm that this will be of infinite service to the king for all navigation in the Indies".

The Spanish government declared as early as 1612 that it would not even consider Galileo's invention, citing as justification that [11/785] negotiations had already begun with a Spanish mathematician for a similar proposal, and therefore "until this is clarified, no new proposals can be accepted". Spain also rejected the Grand Duke's proposed general terms of agreement by replying that the King of Spain would only allow a ship to sail to the Indies if it embarked and disembarked at Seville and was constantly accompanied by a Spanish flotilla.

After four years, in April 1616, Galileo tried to reopen negotiations with Spain.

Belisario Vinta died in 1613; his successor as prime minister was Curzio Picchena, already foreign minister and, like Vinta, a correspondent of Galileo. While in Rome, at the invitation of the ecclesiastical authorities to clarify his position on geocentrism and heliocentrism, Galileo sent a letter to Picchena on April 23, 1616, in which he reported a meeting with Bartolomeo Leonardo d'Argensola, rector of the convent of Villahermosa and secretary to the Spanish count of Lemos. The latter was viceroy (governor) of the Kingdom of Naples, then under Spanish rule. Galileo and Leonardo d'Argensola had discussed various topics, including the measurement of longitude (Appendix A.5). Knowing that d'Argensola would soon return to Naples and from there to Spain, Galileo suggested that they resume the conversation that the Grand Duke had previously encouraged on this subject, assuring Picchena that he would not move without the Grand Duke's approval.

In earlier years, Galileo had realized that the proposal he had made to Spain in 1612 was naive. Although the idea was sound and original, technical improvements were needed that would make it possible to use the telescope on a ship moving over the ocean. He began to study solutions that would allow the observation of Jupiter's satellites from a ship despite the vibrations.

An original arrangement of the telescope mount served to compensate for the ship's vibrations. On a headdress called a *"celatone,"* meaning a large *celata* (helmet), a telescope was attached to one of the holes corresponding to the eyes. With the other eye, free, the observer could detect the light of Jupiter in the sky. The telescope made it possible to see the planet's moons by compensating for the ship's vibrations (Fig. 2.4).

Galileo first mentions the *celatone* in a letter he wrote from Pisa to Picchena in early 1617 (Appendix A.6). In the letter, he speaks of machines he had invented to overcome the difficulties associated with observations from a moving ship; he says that one such device had been made in the arsenal and that he would test it at the first opportunity. Although this machine was different from one in which he had the highest hopes for determining longitude at sea, he wished to have it made because he thought it would be helpful in the Grand Duke's galleys for observing other ships from a distance.

Meanwhile, Galileo continued to improve his measurements of the periods of Jupiter's satellites. Every night, almost without interruption, he was at work observing the satellites. In January 1617, he compiled his so-called table of Bellosguardo, which gives definitive values for the average motions of the planets (third volume of the *Works*). We say "average" because Galileo had noticed some inconsistencies that he could not explain, and he solved the problem by averaging and "adjusting" the values. A physicist in his class was not happy with this and hatched a bug that would haunt him until his later years, when he hired a student to revise his measurements.

He learned that the viceroy had responded positively to his proposals.

On June 30, 1617, he wrote again to Ambassador Orso d'Elci (Appendix A.7), giving more details about his technique and the *celatone*. He felt so confident that he started a deal on the value of the prize he expected from the King of Spain, which included a life annuity of 2000 ducats per year: he wrote that he had heard from Cardinal Borgia that the life annuity was of 6000 ducats, but that he would be satisfied with 4000 (which would become 2000 for his heirs after his death) in addition to the title of Knight of St. Iago. The Order of San Iago (Santiago de Compostela) was the most prestigious monastic-military dynastic order of the time.

Experiments with the apparatus had been carried out with students and later verified by Benedetto Castelli, who had been one of Galileo's favorite students and was now a full professor at Pisa. Castelli informed Galileo [12/1305] that he had

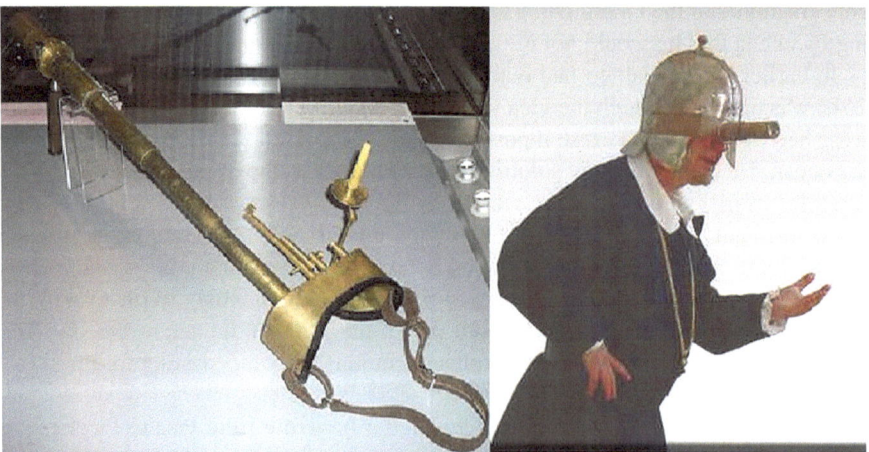

Fig. 2.4 Two reconstructions of the *celatone* based on Galileo's designs. The one on the left is from the seventeenth century and is housed in the Museo Galileo in Florence

> [...] been several times with Signor Giovanni de' Medici, and had shown him, by order of Signor Picchena, the celatone, which had been seen and tried by his Lordship with great pleasure and judged this invention more important than the finding of the same spectacles. [I hope] may at the first opportunity go to Livorno to try some of those young men of whom a choice has been made. [...]

However, Castelli did not consider himself competent to instruct young Tuscan naval officers in using the equipment.

Galileo's letter of May 23, 1618, to Leopold, Archduke of Austria, provides additional information about the *celatone*. Galileo and Leopold had met in Florence when the Archduke had visited the city and the scientist; there was sympathy between the two. After mentioning two telescopes that he had sent to the archduke, Galileo added [12/1324]:

> [...] I am also sending another smaller telescope attached to a brass cover, but this one is made without any ornamentation because it is to serve Your Highness as a model or example from which you can make another that will more accurately fit the shape and size of your head or of anyone else you wish to use it; this instrument cannot be realized without the exact measurement of the individual head and the position of the eyes, as it must be made higher or lower and tilted higher or lower and to the left or right. Surely, Your Highness has no shortage of artisans capable of making an exquisitely crafted headdress based on this model.

In response, the Archduke expressed his gratitude and made his remarks (responding in vernacular Italian).

> [...] I saw the two telescopes and the small cannon with the headpiece, about which the friar Don Benedetto [Castelli], whom I was very glad to see again, informed me in his passage to Pisa. All these things came safely and were found to be true.

Meanwhile, Count d'Elci agreed to reopen the file and suggested that Galileo write two letters on his behalf to the Count of Lemos and the Duke of Lerma. Francisco Sandoval, Duke of Lerma, was the Spanish kingdom's governor (prime minister) and, according to the Venetian ambassador Soranzo, was "as much master of His Majesty's grace and will as no other minister ever was". It must be said that Sandoval and the king were longtime friends: Sandoval had been chosen to be a page at the court and had been close to the king since he was three years old, although at one point he had been mysteriously removed from the court for a few years. He was also an expert navigator, having served as president of the Supreme Council of the Indies.

Galileo prepared the two letters.

Ambassador d'Elci wrote to the Duke of Lerma (Appendix A.8) on September 11, 1617, enclosing Galileo's "general report" of his discovery (Appendix A.3).

The Duke of Lerma moved with his political weight and wrote to the President of the Council of the Indies on November 6, 1617 (Appendix A.9). The Spanish court finally decided to appoint an arbitrator and chose one of the most prestigious: João Baptista Lavanha. Born into a Jewish family in Lisbon around 1555, Lavanha studied mathematics and astronomy at the Roman College. In 1582, King Felipe II recalled him to Madrid as a professor with the specific mission of developing research in cosmography and geography, which was of great use to a maritime power like Spain. In 1586, Lavanha became chief engineer of the kingdom and then, in 1591, royal cosmographer. After returning to Lisbon for a few years to teach the kingdom's naval cadre, he was recalled to Madrid in 1599. Specialized in nautical astronomy and cartography, Lavanha published several important works in these fields. He also worked in shipbuilding until he died in Madrid in 1624.

In reporting this action to Picchena, Count d'Elci could not help but express doubts about the success of Galileo's discovery [12/1286] of November 30, 1617. He raised objections that had probably already been communicated to him by the Council.

> I have seen what Your Lordship writes to me on behalf of Galilei, and I also wanted to read it to Secretary Arostigui, since at the same time that I had the aforementioned letter from Your Lordship, I received an order from the King through the Duke of Lerma that Galilei's offer and proposal should be seen in the Council of State and that I should consult with His Majesty what seemed to the Council.
>
> The secretary had for good that I should make him a brief translation in Castilian of what Your Lordship writes to me, which would excite the Council and give credit to the proposition. I will do all I can to have the offer accepted; but I would very much like the invention to be practicable, and to be always used by all persons, as navigation requires.
>
> By Your Lordship's speech I touch with my hand that by the diversity of the hours in which the same appearance of those stars will be seen around Jupiter, one will at once know the true longitude that those cities or places have between them; but to know this, it's a forcible and necessary thing to see first the aforesaid stars and their aspects, which thing I do not know how will be able to be done at sea, or at least as often and as easily as the necessity of those who sail needs: For, leaving aside that the use of the telescope will not be able to take place in ships because of their motion, but if even there it could have it, it could not and would not serve either by day or in a near time at night that the stars do not appear; and he who sails needs to know hour by hour the degree of longitude in which he is. This is as doubtful as I am burdened with in this matter; and as the difficulty may arise more from my inexperience than from the thing, I will follow to help you warmly, trusting in the good judgment of Your Lordship and in the worthiness of Signor Galilei, who will have thought of everything.
>
> From what the aforesaid secretary Arostigui has told me today, the transaction has already been seen in the Council and the King has been consulted about it, so that His Majesty's decision should soon be known, of which I shall immediately give an account to Your Excellency; and then I shall answer Mr. Galilei.

In response to these objections, which he had probably learned from Picchena, Galileo replied directly to d'Elci on December 26, 1617. He explained that his operation could not always be carried out by everyone in all weather conditions, since it required knowledge of navigation. Moreover, the instrument would be ineffective if it could not be used on a ship, despite the constant movement of the sea. The Tuscan ambassador on the surface expressed his satisfaction with Galileo's reply but expressed concern to Picchena that such limitations on the general application of the invention might diminish its importance.

On January 11, 1618, Orso d'Elci wrote from Madrid to Curzio Picchena [1296], informing him that the Spanish government's decision on Galileo's longitude proposal was still pending: "The King has not even made me answer anything, because the Council of State must first want to inform itself from learned men what it agrees to do."

From d'Elci's subsequent letter to Picchena, dated April 23, 1618 [12/1316], we learn that, in response to his further request of March 13 to the Duke of Lerma, he had been informed that "the proposal had been placed in the hands of some perished men and that he expected their opinion".

The conclusion was not favorable.

In April 1618, Lavanha, probably prejudiced against Galileo, sent his report, which was very negative and full of bias (Appendix A.10). He doubted that Jupiter's satellites existed, even suggesting that they might be an artifact of the telescope; even if they did exist, he doubted that they could be observed from a moving ship; and even if it were possible to observe them from a moving ship, he doubted that the regularity, and therefore calculability, of their rotational motions could be relied upon. In addition, Lavanha said that Galileo's telescope had no special secrets and that the technology to make it was in the public domain.

Based on Lavanha's judgment, the Spanish Council of State later that month, while softening the arbitrator's rudeness, rejected Galileo's idea (Appendix A.11). Nevertheless, Lavanha was asked if he could make a telescope like Galileo's; Lavanha replied no, unless he could get glass lenses as perfect as those that Venice could grind. As a result of Lavanha's answer, however, the council suggested to the king that he have Galileo send him some telescopes.

Galileo waited a year and appealed the decision in 1619 (Appendix A.12). Meanwhile, Giuliano de' Medici, a member of a secondary branch of the grand ducal family, was appointed Tuscan ambassador to Philip III of Spain, succeeding Orso d'Elci, who had spent ten years in Madrid. His post would last until the winter of 1621. The instructions sent to him by Cosimo II in April 1619 advised him to be particularly attentive and to study each character "of that Court, almost all of whom have been changed by a few months here." He was made aware of the rising personalities and political factions at the court, particularly the political influence of the Duke of Lerma, who had been especially sympathetic to the previous Florentine ambassador.

Galileo immediately became interested in the new ambassador, who made a decisive move by allying himself with the Count of Elci and winning the support of the Spanish king's secretariat.

At the end of January 1620, the new resolution of the Spanish Council of State was published, again rejecting Galileo's appeal but requesting that Genoa or Naples receive him with full honors to test his techniques (Appendix A.13).

A few days later, Philip III, King of Spain, wrote [12/1442] to Pedro Tellez y Giron, Duke of Osasuna, Viceroy of Naples (we translate from the Castellan):

Count Orso D'Elci, [then] ambassador of the Grand Duke of Tuscany, on a visit to me, told me that Galileo Galilei, His Highness's mathematician and professor at the University of Pisa offered to give us a way to measure longitude and make oceanic navigation easier and safer, But he said that [the scientist] had not yet been able to experiment with the technique, because he could not come here, and that he also offered another invention for the galleys of the Mediterranean, with which the enemy's ships will be discovered ten times farther than with ordinary sight. I have now received the memorandum, a copy of which is enclosed, in which he asks that the proposal be considered. To know its value, he wants to be sure (as I do) that you will listen to it carefully; and by discussing it with competent people, you will

tell me very carefully what you think of it. With this letter of mine, I tell you that I will feel
well served if you accommodate him and listen to him when he comes to visit you.

As a result, Galileo was invited to Naples in early 1620 to meet with the Spanish
viceroy, who had received specific instructions regarding Galileo's proposal.
Unfortunately, due to Galileo's many hesitations and obstacles and his state of
health, a year passed, and the opportunity was lost.

On February 28, 1621, a series of dramatic events disrupted Galileo's relatively
stable situation and weakened his position. The first is the death at the age of 31 of
the fourth Grand Duke of Tuscany, Cosimo II, who had been Galileo's student and
later his protector. Under Cosimo's strict instructions, the eldest son, Ferdinand II
de' Medici, who became Grand Duke at the age of 11, was to rely on his mother,
Maria Magdalena of Austria, his grandmother, Christina of Lorraine, and a Council
of Regency. The two women, however, did not abide by the agreements and divided
all power between them, beginning a period of crisis in court. In 1628, Ferdinand II
became a grand duke in his own right, but partly because of his mild character, the
influence of his mother and grandmother on public affairs remained strong for years.

A few days later, Philip III of Spain also died; in his place, his 16-year-old son
Philip IV became king. With Philip IV on the throne, the influence of the Duke of
Lerma was reduced to zero due to conspiracies by aristocrats at court. The viceroy
of Naples, Pedro Tellez, the duke's confidant, was dismissed and arrested—he died
in prison.

See the documents in Appendix A for details of the negotiations and related historical events.

2.4   Galileo Proposes His Idea to the States General
of the Netherlands

Even after the second rejection by Spain, Galileo continued to worry about the longitude problem.

In a letter dated October 22, 1627, Alfonso Antonini da Udine, brother of Daniele
Antonini, one of the master's favorite disciples who died prematurely in 1616 while
defending the Republic of Venice, informed Galileo that the Company of Merchants
of the States General of the United Provinces of the Netherlands (provinces that we
will often confuse with the name Holland, which was the most important of all) had
collected and deposited a considerable sum of money to be given to anyone who
could teach a method of determining longitude in navigation. Antonini offered to act
as an intermediary between Galileo and the Dutch. Galileo asked for more information about whom to contact and about the competence of the Dutch in evaluating
scientific matters. He was told that he should communicate with the Lords of the
States General and that the Supreme Magistrate, who governed the United Provinces,
had authority over all matters, including navigation-related matters. The States
General of the Netherlands, formed in 1593, was a parliament that brought together

representatives from each of the seven provinces into which the state was divided. Galileo did not immediately devote himself to the competition called by the States General of Holland.

Favaro mentions the possibility of resuming negotiations with the Spaniards between 1627 and 1629 through the governor of the state of Milan, then under Spanish rule. Still, no documentation is cited in this regard. Probably around the turn of the year 1630, Galileo was preoccupied with the problem of finding a publisher for his *Dialogo sopra i due massimi sistemi del mondo*, his third monograph, the one that would bring him the greatest fame and the greatest misfortune.

Galileo was informed by Gianfrancesco Buonamici, secretary to the Tuscan ambassador to Spain, that King Philip IV wanted to purchase one of his telescopes. He agreed to provide one, pointing out that he had never sold his instruments and had no intention of doing so. He commented that he would have complied with the request sooner if it had come to him directly from the King.

Due to the delay caused by the craftsman who oversaw making the cover for the telescope tube, the instrument could not be completed in time for the diplomatic envoy's departure for Madrid. The device was later transported and delivered, but it was accidentally damaged, and the lenses broke.

However, a rapprochement between Galileo and Spain also seems plausible regarding the longitude problem. Prince Francesco de' Medici, the new ambassador in Madrid, wrote on August 21, 1632, to Andrea Cioli, from 1626 (the year of Curzio Picchena's death) and until 1641 (one year before Galileo's death) Prime Minister of Grand Duke Ferdinand II [16/2290]:

> I beg Your Most Illustrious Lordship to tell Signor Galileo that I have presented to Signor Conte Duca [Don Gaspare De Guzman, Count of Olivares, Duke of Sanlucar, Prime Minister of King Philip IV of Spain (ed.)] the offer he makes of the way of navigating by longitude; and that he may better understand, I have had my secretary translate into this language the writing he gave me at my departure to His Excellency.
>
> Would like the invention to be true, and although he highly esteems the worth of Signor Galileo, yet, as there have been infinite others who have proposed the same, he has some difficulty with it. He told me, however, that he would have everything considered by experts of the profession and would answer me.
>
> He admits that the observations of the motions of the Medicean stars are regular, but he cannot be persuaded as to how one can observe safety in the agitation of the vessel. He approves of everything else and does not mind that the operation remains impeded in turbulent times, knowing that it would be a great purchase even if it served only when serene weather. I will inform your Illustrious Lordship of what will follow, and I will not omit diligence to have a resolution.

This 1632 letter marks the end of the long negotiations between Galileo and Spain.

As is well known, the publication of the *Dialogo sopra i due massimi sistemi del mondo* in Florence in 1632 led to Galileo's immediate summons to Rome by the tribunal of the Inquisition. This was followed by his trial and condemnation in 1633, a condemnation which, after various events, led to his imprisonment in Arcetri.

Galileo returned to the subject of determining longitude in the summer of 1636, nine years after Antonini's letter and three years after his condemnation. He had just completed and delivered to the Dutch publisher Lodewijk Elzevir his last book, the dialogue entitled *Discorsi e dimostrazioni matematiche intorno a due nuove scienze (Discourses and Mathematical Demonstrations on Two New Sciences)*. From this time until April 1640, he conducted negotiations with the States General of Holland mainly through the intermediary of his friend Elia Diodati (1576–1661), a lawyer from Geneva whose family originated from Lucca and who practiced in Paris. The latter was also a correspondent of Constantijn Huygens, secretary to Prince Frederick Henry Stadtholder of Orange and father of the future prominent Dutch astronomer Christian Huygens (1629–1695), who, like Galileo, was also fond of clocks and interested in the problem of longitude.

From Arcetri, Galileo sent on August 15, 1636, to Diodati [16/3341] a formal proposal to the States General of Holland, presenting his solution to the problem of determining longitude. The proposal was contained in three similar letters addressed to three of his Dutch correspondents: one to Martin van den Hove (1605–1639), Latinized as Martinus Hortensius, a well-known Copernican astronomer, optician (he had been a pupil of Snell, who proved the laws of reflection and refraction) and professor of mathematics at the University of Amsterdam; one to Hug de Groot, Latinized as Hugo Grotius or Hugh Grotius (1583–1645), diplomat (Swedish ambassador to Paris), jurist, and historian of the States General; and the last to Admiral Laurens Reael, Latinized as Laurentius Realius or Lorenzo Realio (1583–1637), governor of the Dutch possessions in the East Indies. The letter to Diodati reads:

> I am sending to Your Illustrious Lordship the enclosed letters, all open, and this for two reasons: first, that You may read the whole, thus saving me the trouble of having to reproduce almost every particular contained in them; and then, that You may give grace to place in the inscriptions the names with the titles due to such persons.
>
> While the answers are coming, I will refresh myself a little with silence, taking a break from the work that has long troubled me in the heat of this season, especially to put in order the two works on movement and strength of materials, reduced to dialogues, that I sent six days ago to Mr. Lodewijk Elzevir in Venice, who was about to leave […].
>
> Mister Diodati, my dearest, I am exhausted, for even now I have finished copying the enclosed writings and letters, the work of which, together with composing them, has kept me exhausted for a good four days in this very tedious hot weather.
>
> I finish so much, reserving myself for longer speeches with rested soul and body; and with reverent affection, I kiss your hands.

The accompanying letter to Reael is in Appendix B.1.

Galileo's proposal to the Dutch States General (Appendix B.2) is written in the vernacular, with a summary in Latin.

First, Galileo presented his theory of the motions of Jupiter's satellites, by which experienced astronomers could calculate transits and eclipses. Second, he stated that telescopes of the highest perfection were needed to make these satellites clearly visible and observable. Third, he stated that it was necessary to find a way to overcome the difficulty of using a telescope on board a ship. Finally, he stated that an accurate timekeeping device to indicate the hours and fractions of hours of the solar meridian would be helpful. He wrote that he had constructed a clock that made an error of less than one second [per day].

Compared to the text sent to Spain 20 years earlier, the secrecy and ambiguity had completely disappeared. He went into detail about how exactly longitude could be determined from the appearances of Jupiter's satellites, and about possible measurement errors. The satellites could be observed as they entered Jupiter's shadow cone. Since the dip was very brief, the exact moment could be determined with a margin of error of less than a minute. Another possible method was to observe the conjunctions and separations of the satellites with respect to Jupiter itself. In this case, the observation error would be at most half a minute. A third way was to observe the conjunctions and separations between the satellites themselves, which could be determined with an error of a few seconds at most. These phenomena could be observed every night from anywhere on Earth, at least during the season when Jupiter was visible.

Reael presented Galileo's proposal, accompanied by a Dutch translation, to the Assembly of the States General on November 11, 1636. The new method of determining longitude by observing the eclipses of Jupiter's satellites was described in detail, and it was explained how this method could be used in navigation.

The States General of Holland decided that a commission should evaluate the proposal. Reael, Hortensius, the cartographer Willem Jansz Blaeu (Latinized as Guglielmus Blavius, a pupil of Tycho Brahe and author of the marvelous collection of maps called Atlas Major, now reprinted in Europe by Taschen and classified as Memory of the World by UNESCO) and Jakob van Gool, professor of mathematics and Arabic at Leiden and a diplomat, friend and collaborator of Descartes, were appointed members of the commission. An initial analysis of the proposal took five months, during which the examiners were inundated with letters from Diodati and Galileo demanding a quick resolution.

At a meeting on April 25, 1637, the States General decided to show their gratitude to Galileo [17/3468] by ordering that a gold chain worth 500 florins be brought to him and that Reael be granted 1000 florins to compensate him for his expenses but said they needed further investigation before accepting the proposal.

> Mr. Galileo Galilei, great mathematician, five months have passed since Mr. Reael, former Governor General of the East Indies, offered us as a gift from you the invention of being able to always know longitude, a thing truly desired for many centuries without anyone being able to achieve it until now. We have testified to the aforesaid Mr. Reael that your gift has been very welcome to us, and that we are very grateful for it, knowing also how and when to put our great expense to the test by our most learned mathematicians, I learn and note that I am in these quarters; so that we wait with an unspeakable desire to be clarified by them.

> However, to give you an essay of our gratitude and benevolence, we send you these gifts, accompanied by a gold chain worth about two hundred scudi [500 florins, ed.]; and should your invention prove to be as you have promised us, we shall not fail to express our gratitude more generously, in addition to the honor and reputation that will be due to you by all [translated from French].

Galileo sent Reael a letter detailing his proposal (Appendix B.3) and responding to some of the criticisms that Hortensius had forwarded to him.

This letter is particularly interesting because it illustrates some of Galileo's ideas about the *celatone*. In particular, he suggests that the sailor observing the stars should be placed on a station balanced by a hydrodynamic or oleodynamic suspension, able to reduce the calmness of the observer

to a calmness similar to the tranquility and stillness of the sea. To achieve this benefit, I have thought of placing the observer in a specially prepared location on the ship, where not only the motions from bow to stern but also the lateral ones from side to side are not felt at all. And my idea is based on this principle: if the ship were always in perfectly calm water, with no waves, there is no doubt that using the telescope would be just as easy as on land. Now, I want to place the observer in a small boat within the large ship, with this small boat containing a quantity of water as required, which I will explain shortly. Here, it is evident that the water in the small vessel, even if the large ship tilts or rocks to the right, left, forward, or backward, will always remain level without ever rising or falling in any part, but will always remain parallel to the horizon. So, if we placed another smaller vessel floating in the water within this small boat, it would find itself in perfectly calm waters, and consequently, it would remain without rocking. And this second small vessel is where the observer should be placed. Therefore, I want the first vessel, which must contain the water, to be like a large basin in the shape of a half-spherical orb, and that the smaller vessel should be similar in shape, only smaller, so that between its convex surface and the concave surface of the containing vessel, there is no more space than the thickness of a thumb. This way, a very small amount of water will suffice to support the inner vessel, no less than if it were set afloat in the wide ocean, as I demonstrate in my treatise on floating bodies [referring to *Discourse on Floating Bodies* published in Florence in 1612; see the third volume of the *Works* (Note by the Editor)], which indeed seems marvelous and incredible at first glance. The size of these vessels should be such that the inner and smaller one can support the weight of the person making the observations, along with the seat and other devices necessary for positioning the telescope, without sinking. And to ensure that the inner vessel is always separated from the surface of the containing vessel without ever touching it, so that it cannot be moved by the agitation of the ship, I want some springs to be fixed on the internal concave surface of the containing vessel, or on the convex surface of the contained one, in a number of eight or ten, to prevent the two vessels from coming into contact, but without preventing the inner one from moving with the tilting and rocking of the containing vessel. And if instead of water, we wanted to use oil, it would serve just as well, if not better, and the quantity needed would not be large; two or at most three barrels would suffice.

Note that the problem of observations from a ship had already been investigated by other navigation experts (Fig. 2.5).

He went on to describe the *celatone* in more detail and discuss possible improvements to the project.

[I have made] a certain cap in the form of a helmet which, when worn by the observer and having a telescope fixed to it in such a way that it always looked at the same point that the other free eye was directed towards, without doing anything else, the object viewed with the naked eye would always align with the telescope. A similar device could be constructed, which would not only be worn on the head but also fixed on the shoulders and torso of the observer, with a telescope of the necessary size to clearly discern the small Jovian stars, and it would be so well-adjusted to one of the eyes that by simply directing the sight to the body of Jupiter, the other eye would align it with the telescope, and consequently see the stars close to it.

Fig. 2.5 Perhaps the first illustration of a gimbaled (marine) chair on a ship, designed to allow astronomical observations from a stable position. From J. Besson, *Le Cosmolabe, ou Instrument Universel Concernant Toutes Observations qui se Peuvent Faire*, Paris, Philippe Gaultier de Roville, 1573. Bibliothèque Nationale de France, Paris

He also mentioned that he had made a very accurate pendulum clock with a system of toothed gears that allowed the reading of elapsed time.

The meeting between Galileo and Reael never occurred: Reael died unexpectedly in October 1637, and Hortensius was chosen to replace him. Meanwhile, the Inquisitor of Florence, having learned of Hortensius' expected arrival, had ordered Galileo to do nothing with him because he had been condemned as a heretic. Cardinal Barberini informed the Inquisitor that Galileo would not even be able to accept the necklace, so if it reached him, Galileo would have to return it. The Inquisitor's intervention was useless, however, because while he was planning for his trip to Florence, Hortensius also died. The other two examining committee members, Willem Blaeu and Jakob van Gool, had also recently died. A curse had caused all the members of the commission that was to evaluate Galileo's idea to die in less than two years, and the evaluation process had to start all over again.

Although now in feeble health, Galileo seems to have held on to hope until the end. His last letter on the subject is dated January 15, 1640 (Appendix B.5). He regretted the death of the four commissioners, expressing his grief in a way that makes us reflect on his pugnacious character:

> I am deeply sorry for the sudden death of Mr. Martinus Hortensius, which occurred after the deaths of the other three Commissioners. These misfortunes, added to my own injury, seem to be hindering and disrupting the continuation and progress of the project I have undertaken with the Most Illustrious and Most Powerful Lords States. [...]

and proposed to send his own pupil Vincenzo Renieri to Holland to negotiate on his behalf.

The business, however, as far as it will be possible for me, will not remain impeded or delayed, since I have met a person most intelligent in these astronomical sciences and most apt to not only make up for the defect caused to me by blindness, but to carry it on with greater accuracy, being, in addition to perfect intelligence, a man of nimble and shrewd wit, of complexity and freshness of age fit for all labor, of keenest sight, of an ardent desire to go forward, embracing the enterprise with great enthusiasm, and (what pleased me most) even willing to move to Holland.

The young (he was born in 1606) Benedictine monk Renieri, who would soon ascend to the chair of mathematics in Pisa in 1640, was studying the motions of Jupiter's moons for Galileo. Galileo had asked him to update and verify Bellosguardo's table, but the work was not completed until after Galileo's death.

In a moving letter to Galileo dated June 15, 1640, which is the tragic epilogue to this long and troubled affair (Appendix B.6), Diodati for the first time openly expresses his opinion that hope for further progress is vain.

Among his acquaintances and friends, only Fulgenzio Micanzio of Venice, a willing assistant of Fra' Paolo Sarpi and collaborator, proofreader, and, so to speak, literary agent (he had conducted part of the negotiations with the publisher) of Galileo's last book, that dialogue entitled *Discorsi e dimostrazioni matematiche intorno a due nuove scienze,* published in Leiden in 1638, remained to encourage him to the last. In a letter dated January 4, 1642, Micanzio informed Galileo of the desire of some Dutch gentlemen and merchants to see the realization of his work, so extraordinary and considered impenetrable by the most eminent intellectuals.

This letter would never reach Galileo: on January 8 of that year, the Tuscan genius died in his Villa Gioiello in Arcetri. But the geolocation technique by means of Jupiter's satellites did not die, as we shall see.

One of Galileo's favorite students (along with Benedetto Castelli, Paolo Aproino, and Daniele Antonini in the Paduan period and Evangelista Torricelli in the Tuscan period), the Florentine Vincenzo Viviani, in the biography *Racconto istorico della vita di Galileo Galilei*, published in 1654, summarized the affair as follows.

[In 1636,] he decided to make a free offer to the Most Illustrious and Powerful States General of the United Provences of the Netherlands of his admirable find for the use of longitudes, with the patronage of Signor Ugo Grozio, Ambassador resident in Paris for the Majesty of the Queen of Sweden, and with the most ardent employment of the aforementioned Mister Elia Deodati, through whose hands all the negotiations then passed.

Such a generous offer was eagerly received by the States, and in the course of the treaty, it was adorned with their most humane letter, accompanied by a magnificent gold necklace, with which Signor Galileo did not wish to decorate himself at that time, They begged the States to be pleased that their gift should remain in other hands until the enterprise was completed, lest they should give matter to his malignant emulators to pass him off, by vain oblations and presumptuous notions, as the expropriator of the treasures of great lords. They also intended to give him the most significant recognition in the event of a fortunate success. They had already appointed for the examination and experience of the proposal four Referees, principal mathematicians, experts in navigation, geography, and astronomy, to whom then Signor Galileo freely confided all his thoughts and secrets concerning the speculative and practical nature of his invention, and in addition, all the artifices he imagined to reduce, if necessary, to greater ease and safety the use of the telescope in the mediocre agitations of the ship for the observation of the Medicean stars. It was examined by the Referees and they, with admiration, approved such an ingenious proposal. It was chosen by the same States Mr. Hortensius, one of the four Referees, to move from Holland to Tuscany and to talk with Mr. Galileo to extract all those documents and instructions from his voice, on the theoretical and practical aspects of the invention. In short, in the continuation for more than five years of this contract, it was not for one or the other of the parties to take the necessary precautions and decisions to conclude such an undertaking.

But the Divine Will, not yet satisfied, was well pleased that our Galileo should be recognized as the first and only discoverer of this so longed-for invention, as well as of all celestial novelties and wonders, and that for this he should make himself immortal and praiseworthy together with the earth, the sea, and almost, I say, with heaven itself; But he willed, by various accidents, to prevent the execution of the enterprise, and to postpone it to other times, while he suppressed the sumptuous pride of men, who, by this means, with equal safety, might the unknown ways of the ocean, as the most learned of the Earth.

Wherefore, Mr. Galileo, for the space of twenty-seven years, suffered great inconveniences and labors to correct the motions of Jupiter's satellites, which he had finally obtained with supreme adequacy for the use of longitudes. More, having for exact observations some years before, and before any other, perceived with the telescope a new motion in the lunar body using its spots; not permitting the same Divine Providence that a single Galileo should unveil all the secrets which perhaps by the exercise of the future life it keeps ascending in the heavens; in the greatest heat of this treatise, at the age of about seventy-four years, visited him with a most troublesome fluxion in his eyes, and after a few months of troublesome infirmity deprived him of all those who alone, and in less time than a year, had discovered, observed, and taught him to see in the universe much more than had been allowed to all human views put together in all the centuries past. By this merciful accident, he was compelled to place all his own writings, observations, and calculations concerning the said planets in the hands of his pupil, P. D. Vincenzo Renieri, who was later to become a mathematician of Pisa, so that he might make tables and ephemerides of them, compensating for his blindness, and then give them to the States and communicate them to Signor Ortensio, who was to appear here.

But in a short time, not only the death of this one, but also of the other three commissioners who had been assigned to such handling, came to be known, to be fully instructed and assured of the truth of the proposal and of the safety and manner of its execution. And finally, when by Mr. Huygens, First Councillor and Secretary of the Lord Prince of Oranges, and by Mr. Borelius, Councillor and Pensioner of the City of Amsterdam, personages of the clearest fame and literature, they were incessantly procuring to summarize and perfect the negotiations with the same States; and that Mr. Galileo had decided, with their consent, to send with them Fr. Vincenzio Renieri, as most informed of every secret, with the tables and ephemerides of the Medicean planets, to confer the whole thing and to instruct anyone whom they pleased; when I say, from these, who already learned the proposal for infallible and of most certain event, this was negotiated with every greater enthusiasm, the author of so great an invention, as I shall say below, lost his life: and here all treaty with the States of Holland was totally severed.

[...] But let the practice of such a noble discovery be postponed, for whatever reason, and let others endeavor to trace the movements of those stars with their own sweat, or let others, adorning themselves with the labors of the first discoverer, attempt to make themselves the author of them, to extract from them rewards and honors; for just as for the determination of longitudes the means of Jupiter's companions is the only one in nature, and therefore this alone will one day be practiced by all observers of land and sea, so the primacy and glory of the invention will always be that of our great Galileo, authenticated by whole kingdoms and the most illustrious republics of Europe, and to him alone the correction of nautical and geographical charts and the most accurate description of the entire globe will always be due.

Note Viviani's prescient certainty that in the future, geolocation would use the movements of Jupiter's satellites "or others."

2.5   The Use of Jupiter's Satellites in Cartography

The method of transit of Jupiter's moons, technically impractical at sea, proved valuable and accurate for ground-based measurements.

Gian Domenico Cassini, a professor in Bologna, calculated the tables of motions of Jupiter's four satellites more precisely than Galileo and published them in 1668 in his *Ephemerides bononienses mediceorum siderum.*

Gian Domenico Cassini (also known as Cassini I, 1625–1712) would greatly influence the development of astronomy in France. After being appointed full professor in Bologna in 1650, he was elected director of the Observatoire de Paris in 1671. His son, Jacques Cassini, was appointed director, as did his grandson, César-Francois Cassini, an astronomer and geodesist. Great-grandson Jean-Dominique Cassini, nicknamed Cassini IV, would be the last of the Cassinis to hold the post of director of the Observatory, a position he held from 1784 until the French Revolution.

The ancestor of the Cassinis is also remembered for having drawn the so-called Paris meridian, which runs through the center of the Observatoire de Paris (whose main observation room is now called the Cassini Room). It was not until 1911 that France officially replaced the Paris meridian with the Greenwich meridian as the zero meridian (ratifying, twenty-seven years late, a decision made by the international committee that had chosen Greenwich).

Also, thanks to Cassini's influence, the technique of Jupiter's moons (in particular Io) became the official method for determining longitude in France. The moons of Jupiter were used for the French Académie des Sciences project to draw a new map of France in 1684. The technique showed that the coastline extended much farther east than on previous maps, and the size of the country's surface was significantly reduced. Louis XIV commented that the astronomers had stolen more land from France than he had won in all his wars (Fig. 2.6). Despite the king's displeasure, Galileo's technique remained the reference for the French for several decades.

By the end of the century, many had seen Jupiter's satellites and studied their motions (Fig. 2.7), and no one doubted their existence or the regularity of their orbits.

The abbot and astronomer Jean Picard, a former assistant of Gian Domenico Cassini, also used Galileo's method to improve the cartography of the island of Hven in Scania. On this island, Tycho Brahe's old observatory was located.

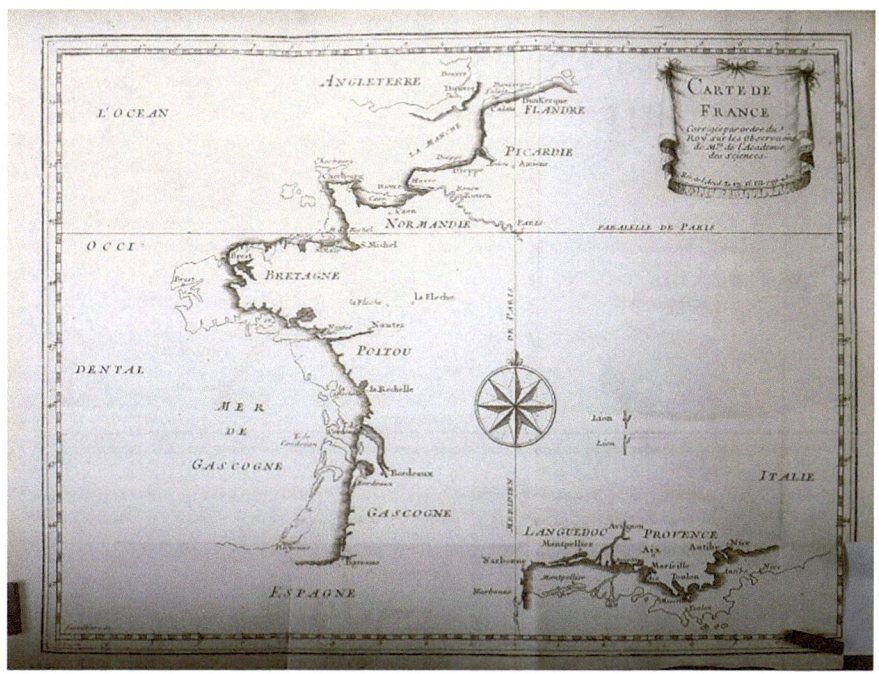

Fig. 2.6 Map of France presented to the Académie des Sciences in 1684, showing the outline of an earlier map (flat) compared to the new survey (heavier outline). Académie des Sciences, Paris, public domain

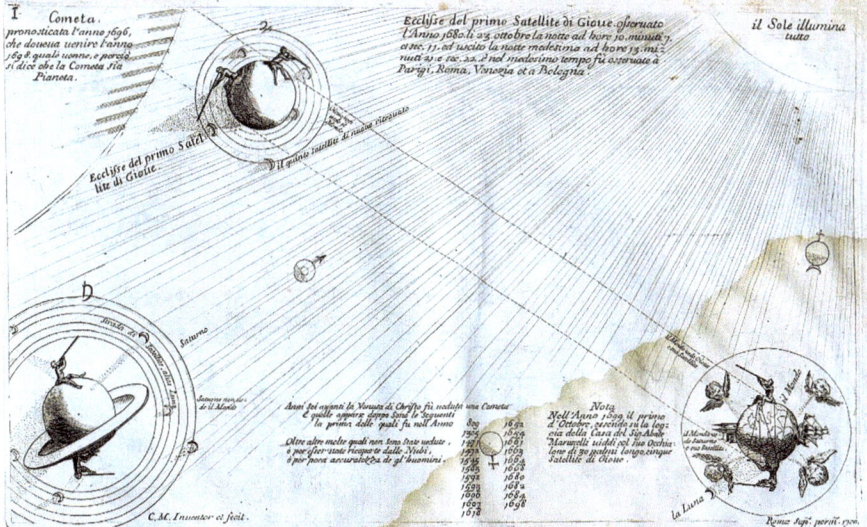

Fig. 2.7 Jupiter's satellites (note that five are erroneously depicted, while only four were known at the time) from the book *The Art of Making Rivers Navigable in Various Ways, with New Inventions and Various Other Secrets* by Cornelius Meyer (Rome 1696), ETH Zurich Library

Chapter 3
From Galileo to Today

3.1 The British Competition and Its Outcome

In the eighteenth century, the methods for astronomical determination of lunar transits were improved. At the cost of complex calculations, an accuracy of about one degree was achieved.

As we have seen, combining an accurate clock with simple solar observations could have solved the longitude problem. Galileo also moved in this direction in his later years without achieving the necessary precision. The Dutch physicist Christiaan Huygens (1629–1695) made significant improvements in the technology of pendulum clocks, and in 1657, he patented a model that was more accurate than earlier mechanical clocks. However, pendulum clocks did not withstand the motion of a ship well enough: after a series of tests, he concluded that other solutions were needed. Later, Huygens was the first to use a balance spring as an oscillator in a clock, which made it possible to create accurate portable clocks, but still not enough to measure longitude with enough accuracy.

In 1714, the British Parliament launched a competition to encourage the discovery of a reliable method for determining longitude at sea. By an Act of Parliament, it offered a reward of £20,000 (a very substantial sum at the time, equivalent to nearly four million euros today according to the Bank of England's inflation calculator) to the first person who could develop a reliable method of determining longitude to an accuracy of better than half a degree (about 55 km), and half that amount to anyone who could achieve an accuracy of one degree. The Earth rotates one degree of longitude every four minutes, so the maximum tolerable error in measuring time was a few seconds per day; at that time there were no clocks capable of achieving such accuracy on a moving ship.

The solution was eventually found by John Harrison, a self-taught English watchmaker who devoted much of his life to developing a marine chronometer that could solve the problem. Over thirty years, Harrison created several models of increasingly accurate marine chronometers. His fourth and most famous model was finally accepted as a reliable method for determining longitude. Despite opposition and initial rejection from the scientific authorities of the day, Harrison received

A. De Angelis, *Galileo and Satellite Navigation*, SpringerBriefs in History of Science and Technology, https://doi.org/10.1007/978-3-031-78799-7_3

significant support from the British Navy and Parliament. In 1773, at the age of 80, after an appeal against the British government's decision to cancel the prize, Harrison finally received compensation that was reduced to £8750. Harrison's story is narrated in Dava Sobel's excellent book *Longitude*.

Harrison's discovery significantly impacted maritime navigation, allowing for greater safety and accuracy during ocean voyages. Harrison's marine chronometers paved the way for the development of even more advanced technologies for navigation and represent an important milestone in the history of maritime science and navigation.

In the early years, chronometers were very expensive. So, thanks in part to improvements and simplifications in lunar almanacs, the lunar transit method was even more common than Harrison's until the 1800s.

The cost disadvantage diminished as chronometers began to be mass-produced. New watchmakers simplified the design and manufacture of chronometers. Between 1800 and 1850, as chronometers became more affordable and reliable, they gradually replaced the lunar distance method. By 1850, the vast majority of the world's mariners had abandoned the lunar distance method.

However, until the early twentieth century, experienced mariners continued to learn and sometimes use the lunar distance technique. Only in 1907 did British naval officers stop learning to use this technique.

Everything changed with the discovery of electromagnetic waves and the possibility of wireless signal transmission.

3.2   Electromagnetic Wave Communications Before the Space Age

In 1897, Guglielmo Marconi, aged twenty-three, patented wireless telegraphy, which he had successfully experimented with two years earlier. In 1901, he sent the first transatlantic electromagnetic signal from England to Canada. The potential of using "wireless" signals to determine location became apparent long before Marconi was awarded the Nobel Prize in Physics for his discovery in 1909. Wireless telegraphy was used to extend and refine the telegraph network, providing greater accuracy than any clock in determining the time of the reference meridian (and thus longitude) in the oceans.

The transmission of wireless time signals for use on ships at sea began in 1907 from Halifax, Nova Scotia. In 1910, time signals were transmitted from the Eiffel Tower in Paris. These signals allowed navigators to check and adjust their clocks. The Second International Radiotelegraphic Convention, held in London in July 1912, established international standards for maritime radio communications to provide interference-free time signals worldwide. The sinking of the Titanic only three months earlier was a major factor in speeding up the diplomats' deliberations.

Land-based observers, especially cartographers and explorers, also used wireless time signals.

A few years later, aircraft began to take to the skies. For localization, pilots relied mainly on celestial navigation, calculation by approximation from certain points, and visual references, with very few instrument resources. Hard as it may be to believe today, until less than a century ago, celestial navigation was the standard method for determining an aircraft's position. Navigators used a sextant to calculate the aircraft's position relative to the Sun, Moon, or stars. This method was used until the jet age of the 1960s, and early versions of the Boeing 747 (the famous jumbo jet) in 1969 even had a sextant hatch on the cockpit roof.

Calculation by approximation was another standard method of long-range navigation. In this technique, navigators used previously known positions to estimate the aircraft's current position based on speed and flight time. Although weather conditions could affect these estimates, it was still a relatively accurate way to calculate the aircraft's position.

However, more accurate navigation became essential with the use of aircraft for military purposes, which involved flying at higher altitudes and over greater distances.

After World War II, several navigation technologies and methods were developed that enabled the development of sea and air travel. These include techniques based on determining the distance of the aircraft relative to fixed points, determining speed relative to these points, and using the emission and detection of electromagnetic waves (radiolocation) based on principles explored by Guglielmo Marconi. These techniques are completely independent of astronomical observations.

Radiolocation or radionavigation exploited radio signals transmitted by ground stations or, more generally, by radio stations placed in known locations. Different stations emitted signals of different shapes; users could use various techniques to associate each signal with the station that had emitted it: the shape of the signal itself, the intensity and direction, and the frequency. These could be used to measure the position (and also speed, taking advantage of the change in frequency associated with the so-called Doppler effect, which occurs when the source of a wave and the receiver are in relative motion). The redundancy of the information made it possible to compensate for any errors.

But by the end of the twentieth century, a new revolution was underway, linked to the beginning of the space age. The localization technique by satellite positions, which Galileo had envisioned three hundred and fifty years earlier, finally became feasible thanks to artificial satellites.

3.3   Satellite Location and the Legacy of Galileo

On October 4, 1957, the Soviet Union launched the first artificial satellite, Sputnik 1. The launch of Sputnik sparked a reaction of surprise and concern in the Western world, particularly in the United States, which was prompted to engage with great energy in the space race. The idea of exploring space and reaching new horizons captured the imagination of people, especially young, and influenced art, literature, and popular culture. The launch of Sputnik also revolutionized technological development and education. There was a strong global push toward science and technology, with large investments in scientific research and education in mathematics, science, and engineering.

In 1958, the U.S. founded its own space agency, the National Aeronautics and Space Administration (NASA), and began a series of space missions to try to catch up with the Soviet Union in terms of technology.

Despite Cold War tensions, the United States and the Soviet Union also undertook forms of collaboration in space exploration. In 1975, the Apollo-Soyuz project marked the first joint flight between U.S. and Soviet astronauts, symbolizing a moment of détente and again capturing the imagination and enthusiasm of people, particularly young people.

But let us now come to how artificial satellites made Galileo's idea feasible. The weakness of Galileo's technique was related to the fact that Jupiter's satellites visible to the naked eye are only four, quite distant from Earth (between six hundred thousand and nine hundred thousand kilometers) and angularly close together— these two features limit the accuracy of targeting. In addition, the technique worked only at night.

With the advent of artificial satellites, it was possible to launch constellations of satellites at the desired distance and in such numbers that a large proportion of them is always visible from anywhere on the globe. Satellite signals could be transmitted by electromagnetic waves detectable day and night.

Localization is done through the triangulation technique, which we describe in brief.

In triangulation, which geometers have used since ancient Greece, the position of an object is identified by distances from three reference points on the Earth's surface. Let us take an example: if we know that we are 651 m away from my house, which is located on Via Patriarcato in Padua; 454 m away from Galileo's house, which is located on the street now called Galileo Galilei; and 727 m away from my office, which is located on Via Belzoni, then we can conclude that we are in the Bo building, the headquarters of the University of Padua and the rectoral offices (see Fig. 3.1).

The same basic idea of triangulation can be applied to distances from artificial satellites whose position is known (see Fig. 3.2). However, this time we are dealing with intersections of spheres rather than circles. This is the principle of satellite geolocation.

The United States came before the Soviet Union in the race for satellite geolocation thanks to the Global Positioning System (GPS). GPS is one of the global navigation satellite systems (GNSS) that provide valuable information for geolocation and the exact time to receive devices placed anywhere on or near Earth if they have a clear line of sight to at least four satellites (we will see shortly why four satellites are used and not three) in the constellation. It does not require the user to transmit data and is independent of telephone or Internet reception. The system provides

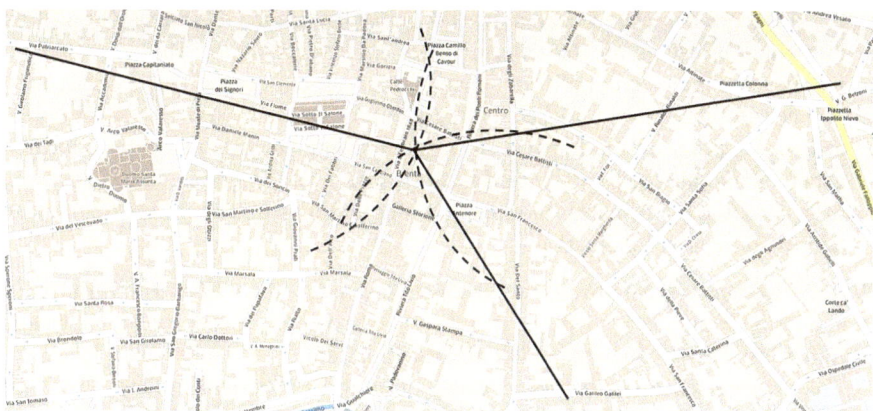

Fig. 3.1 Measurement of the position of the Bo Palace in Padua by triangulation on the Earth's surface from my house (left), my office (right), and Galileo's house (bottom)

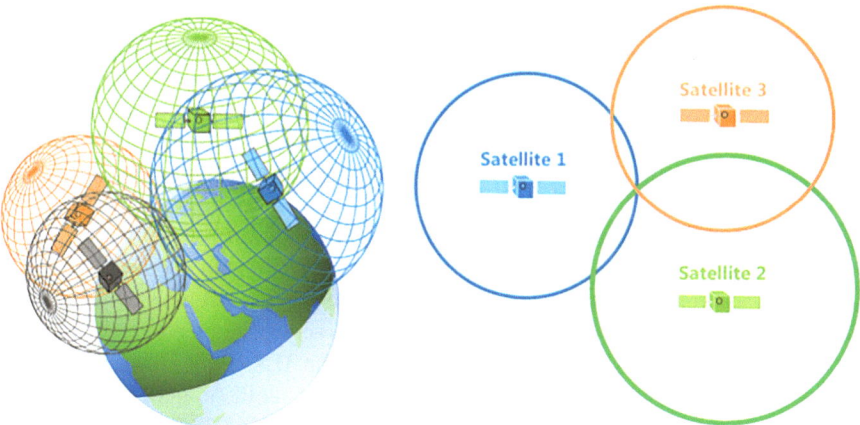

Fig. 3.2 Measuring the position of a point by triangulation from space

essential positioning capabilities for military, civilian, and commercial users worldwide. GPS is freely accessible to anyone with a suitable receiver.

The way a GNSS system works can be summarized as follows:

- Satellites constantly send out radio signals containing information about their position; from these signals, receivers can determine which satellite sent the signal (each satellite sings a different song, so to speak) and the exact time it sent the signal.
- A receiver, such as a GNSS-enabled cell phone, receives satellite signals.
- The GNSS receiver measures the time it takes for the signals to reach it from each satellite. Since the signal speed (the speed of light) is known, the receiver can calculate the distance between itself and each satellite.

– For distances calculated from at least four satellites, the GNSS receiver uses an interpolation technique to determine its precise position. The interpolation involves the intersection of four spheres, each representing a distance measured by one satellite. As we have seen, three satellites would be sufficient in principle to locate a point on the Earth's surface, but redundancy dramatically improves accuracy and allows for estimated position error. The location can be expressed in geographic coordinates (latitude and longitude) or other formats, such as an address or nearby landmarks.

Why do we need signals from four satellites instead of three? Let us return to our triangulation example. We will see that there are two good reasons, one obvious and the other less easy to understand but no less important.

The obvious reason is that the intersection of three spherical surfaces (Fig. 3.2) defines two points, which, by adding uncertainty about the initial time, become an arc rather than a point. This means that it can define a point on the Earth's surface, but if I also want to know the height of that point relative to sea level, I need a fourth satellite.

The main reason is the accuracy of GNSS receivers. To travel from a satellite to the receiver, the signal needs a time interval roughly equal to 20,000 km (the typical satellite altitude) divided by three hundred thousand kilometers per second, or the speed of light: about seven-hundredths of a second. If I want to locate my position within 10 m, I need to measure these seven hundredths to within thirty billionths of a second. Only an atomic clock, which costs tens of thousands of dollars, is accurate enough. The normal clocks in our mobile phones, which cost about thirty dollars, have an accuracy of ten-millionths of a second. With a fourth satellite, the desired accuracy can be achieved by recalibrating any systematic errors in our quartz clock.

GNSS receivers constantly update their position based on signals received from satellites. This enables users to perform real-time tracking and navigation.

After five years of study, the first satellite in the U.S. GPS constellation was launched in 1978, and the whole constellation of 24 satellites, making it a global system, became operational in 1993. Since 1994, the use of GPS has been authorized for air navigation.

In general, the use of GPS provides positioning with an accuracy of five meters.

Beginning in the early 1990s, the location accuracy provided by GPS was improved by the U.S. government through a program called "selective availability," which could selectively degrade or deny access to the system at any time, as happened to the Indian Army in 1999 during the Kargil War. The Russian global navigation satellite system is called GLONASS (transliterated from Russian as Globalnaja Navigacionnaja Sputnikovaja Sistema, in English GLObal NAvigation Satellite System) and was developed shortly after GPS (it has been operational since 1995) but suffered from incomplete coverage of the Earth until 2000. It, too, has been made available to all humankind. GLONASS and GPS reception can be combined in a single receiver, allowing for faster position estimates and greater accuracy, typically about 2 m.

But it was unacceptable for the rest of the world the United States and Russia to be the only owners of the geolocation systems, since they could have blackmailed the planet's inhabitants at any time. As a result, in recent years, several countries have created or are creating other global or regional satellite navigation systems.

China started launching its BeiDou global satellite navigation system in 2018 and completed its deployment in 2020. Beidou has comparable accuracy to Russian and U.S. systems.

The European Space Agency (ESA) launched the Galileo global satellite navigation system (Fig. 3.3) in 2005, made it available in 2016, and completed it in 2020 on behalf of the European Union. Galileo provides slightly better localization than the Russian, U.S., and Chinese systems due to a more accurate internal clock. In addition, Galileo is neutral and available to all countries and everyone. A new generation of Galileo satellites will become operational after 2025 to replace the current constellation, which will be used as a backup.

India and Japan are developing satellite navigation systems, but these are currently focused primarily on their respective countries and neighboring regions and therefore cannot be considered global. Bear in mind that the cost of a GNSS system is around thirty billion dollars: the technology is beyond most individual countries' reach.

Satellites used for geolocation are placed in so-called MEO (Medium Earth Orbit) orbits, between twelve and thirty thousand kilometers above the Earth's surface. Satellites in MEO orbit spend more time above the horizon than those in LEO

Fig. 3.3 A satellite of the Galileo constellation (credit: ESA)

(Low Earth Orbit) orbit, at an altitude of about five hundred kilometers, where most satellites, such as the International Space Station, orbit.

A single satellite in MEO orbit is visible for longer periods, allowing for better visibility from devices on the Earth's surface. MEO orbits also require fewer satellites than LEO orbits to provide global coverage, reducing the costs of launching and operating satellites.

Finally, it is important to note that navigation systems need to see satellites to receive signals accurately. Obstacles such as tall buildings, dense trees, or unique urban environments with narrow streets (those who live in Venice know this well) can affect signal reception and reduce the accuracy of positioning. In addition, GNSS has poor reception inside buildings.

3.4   Ideas for the Near Future

Due to the attenuation caused by building materials (GPS satellites transmit in the so-called L-band of microwaves, corresponding to wavelengths of about twenty centimeters), the positioning signal from GNSS satellites loses significant power inside buildings, reducing the possibility to get in touch with at least four satellites. In addition, reflections from walls introduce errors in distance measurements. Today's GNSS receivers are becoming more sensitive due to the increased processing power of microchips. High-sensitivity GNSS receivers can receive satellite signals in most environments and attempts to determine three-dimensional indoor position are successful.

A promising direction for error correction in indoor localization is to use alternative sources of navigation information, such as Wi-Fi transmission systems, which can be integrated with GNSS signals through hybrid localization algorithms.

The current goal for indoor systems is to achieve accuracies on the order of 30 cm, which will enable orientation in hospitals and other complex structures, improve our experience in video games, and be an indispensable aid for autonomous ground and drone transportation.

Postscript

The idea to write this book was given to me at different times by two colleagues, whom I thank very much for allowing me to study, learn, and tell (three of the things I enjoy most in life). The first colleague is Patrizia Tavella, Director of the Time Department of the Bureau International des Poids et Mesures in Paris, the office that calculates the Coordinated Universal Time distributed throughout the Earth—every time I see that number scrolling on a somewhat old-fashioned screen in Sèvres, near Paris, I can't help but think that we Earthlings need to realize as soon as possible that we are one and the same and that we should have a common consciousness of the planet. The second is Alessandra Fiumara, who is in charge of scientific relations between Italy and ESA, and who was looking for interesting topics on geolocation for a conference she was planning to organize.

Michele Camerota provided me with bibliographical sources and the update of the famous national edition of Galileo's works edited by Antonio Favaro, a reference for any study of the Tuscan genius. Flavia Marcacci was helpful in my search for bibliographical material, and Bill Shea was generous with advice, and talking to him is always a pleasure. On top of his suggestions, he encouraged me to write this text in English.

Giovanni Busetto, Michela De Maria, Alessandra Fiumara, and Bill Shea read the text and gave me valuable suggestions.

Appendix A: Documents Related to the Presentation of Galileo's Proposal to the Kingdom of Spain

[11/757] Belisario Vinta, Prime Minister of Tuscany, to Orso d'Elci, Ambassador to Madrid (Florence, September 7, 1612)

His Serene Highness [the Grand Duke of Tuscany] wishes to inform His Majesty [the King of Spain] of a new invention that could bring the final touch of perfection to navigation by providing a method to measure longitude at any hour of the night and almost throughout the entire year. This method was recently discovered by Galileo Galilei, a vassal of His Highness, and his philosopher and chief mathematician. Galileo is the same scientist who, through his telescope—or as we might say, through his long-distance eyeglasses—has uncovered numerous celestial phenomena and motions that were unknown to all our ancestors. He has demonstrated these discoveries many times to Their Highnesses and to experts in this field across Italy, earning such credibility that we have no reason to doubt the accuracy of his findings. The essence of this method lies in a new celestial discovery of Galileo's, previously undecipherable, which resolves the mysteries that perplexed earlier astronomers and geographers who could not reach such a conclusion.

If it pleases His Majesty to consider this, His Serene Highness will instruct Galileo to prepare a detailed report on all aspects of this method, which will be sent to His Majesty for his consideration and final judgment.

[5/p419] The Proposal of Longitude by Galileo (1612)

That supreme and marvelous problem of finding the longitude of a given place above the surface of the Earth, so longed for in all past centuries, because of the very important consequences which depend on such a finding in geography and nautical charts, and their complete perfection, has excited the toil of various minds up to the present age, not only to restore to it that glory which such an invention may

A. De Angelis, *Galileo and Satellite Navigation*, SpringerBriefs in History of Science and Technology, https://doi.org/10.1007/978-3-031-78799-7

deservedly claim, but also to obtain again the real rewards and remunerations proposed to the inventor. But hitherto all efforts have been in vain, nor have ever been able to make greater progress than that which was found by the ancients, and especially by Ptolemy, with subtle and noble invention; and perhaps it would have been absolutely impossible to solve such a problem, if other stupendous problems had not first been found by human ingenuity, and at first glance much more difficult to solve than the same problem of finding longitude.

And for better explication, I will briefly recapitulate what is longitude and latitude of a certain place above the surface of the Earth, and how that has been found so far by the ancients, and in how many difficulties wrapped and tangled.

Latitude is the arc of the meridian taken between the apex of a point and the equinox, which arc is always equal to the arc of the same meridian taken between the pole of the world and the horizon, that is, to the elevation of the pole of that place. The longitude, then, is nothing but the arc of the equinoctial line drawn between the meridian of one place and the meridian of another: and as it has been generally established by cosmographers that the meridian passing through the Canary Islands is the prime meridian, it will be said that the longitude of a place is the arc of the equinoctial line drawn between the meridian passing through the Canary Islands and the meridian of the place.

Now it must be known that all the ways of finding such a longitude hitherto proposed have been justly recognized as vain and fallacious, from two onward.

The first would derive from the knowledge of the travel itinerary between the place and the prime meridian; but such a procedure remains totally useless, if between the two meridians is interposed some vast sea, or other tract of space impassable by path.

The other way, hitherto used by the great cosmographers, is by means of lunar eclipses. This is the most exquisite way that has yet been practiced: nevertheless, it still suffers from many and very serious difficulties. And to explain them briefly and as simply as possible, let it be, for example, that the longitude of Rome is sought for a lunar eclipse to be seen in Rome on December 20, 1638, at 1:30 p.m., and the same eclipse in the Canary Islands at 11 a.m.: it is manifest that the meridian of Rome is found to be more easterly than that of the Canary Islands for two and a half hours; and as one hour imports 15 degrees of equinoctial, we will say that the longitude of Rome is 37 degrees and 30 minutes.

Now, as has been said, this method of determining longitude is subject to several difficulties. The first is the rarity of lunar eclipses: there are no more than two visible lunar eclipses per year, and sometimes only one, and sometimes none. Moreover, it is very difficult to observe precisely the beginning or the middle or the end of an eclipse; so that when the Moon begins to plunge into the cone of the Earth's shadow, here the shadow is so faint and blurred that the observer is puzzled whether the Moon has begun to be eclipsed or not.

I do not think it can remain in doubt to anyone who understands these things, that if a way were found to make use of more frequent eclipses, so that where we have so few at the top of the year that you can say that if only one were made, we could have three or four or five and even six per night, this would imply a very great advantage: then there would be more than a thousand such eclipses a year. And if they were not really lunar eclipses, but phenomena in the sky and appearances equivalent and similar to lunar eclipses, it is manifest that the gain would be very great. Moreover, since, as has been said, lunar eclipses are precisely unobservable in their principles, means, and ends, so that one can perhaps err by more than a quarter of an hour (which would be an error in longitude of about four degrees), it is manifest that if the measurement is reduced to such exquisiteness that one does not err by a minute of an hour, one would still have made a progress of the greatest consideration. Moreover, the tables of the motions of the Sun and Moon, on which the calculation of lunar eclipses depends, are not yet reduced to such exquisiteness that we do not err by a quarter of an hour, and perhaps more; so that, if we had to make use of said tables, we might make an error in longitude of about eight degrees: and therefore it is manifest that if our eclipses, or whatever other phenomena there may be, were dependent upon and regulated by such exquisite tables, that there would not be an error of one minute, the whole business would be, it may be said, reduced to perfect perfection, as far as our knowledge can reach.

Now I say that the great ingenuity and labors of Signor Galileo Galilei, chief philosopher of the Most Serene Grand Duke of Tuscany (to whom Signor Galileo deservedly owes the title of Great), have come to discover in the heavens things totally unknown to past centuries, equivalent to more than a thousand lunar eclipses every year, observable with minute precision, and, what is most important, reduced to calculations and righteous and exquisite tables.

And this invention would be submitted to the King's great Majesty, with the request that, if for any reason such an offer should not be accepted, His Majesty would be graciously inclined to grant grace, that if in future times others more fortunate should present the same enterprise and it should be accepted, not for this reason should Mr. Galileo or his descendants be deprived of those honors and graces which were intended for the inventor himself by the greatness of the royal benevolence.

It is true that this proposal, at the first look, may perhaps appear an absolutely impossible paradox, and unworthy of being heard: with all that, it does not seem that the importance of such a noble undertaking deserves to be condemned to a vanity, if it is not first carefully examined and considered by intelligent persons of the profession. It must also be considered that, to reduce what is proposed to practice, it is necessary to distinguish it into parts, of which some belong absolutely to Signor Galileo, and others seek the royal magnitudes and powers.

It is up to Signor Galileo to show the way of working, to warn of the diligence that is required, to present in enlarged form all the tables that we need, and to propose all that is necessary to achieve our purpose. But, on the other hand, as we are dealing with a multitude of men to be employed, and first to be instructed and disciplined, and as it is more necessary to navigate with large and strong ships over vast seas, and as it is necessary for the instruction of men to establish academies, all things that cannot depend but on the greatness of monarchs and great kings, this part is not to be desired or sought by the tenuous fortune of Signor Galileo, but by the orders of His Majesty, as will be more minutely set forth when the occasion arises.

Nor should a most important consideration be omitted: namely, that in proposing this enterprise anew, with new sciences and arts, still that everything is proposed (as will be seen) by means already reduced to a high degree of perfection, with all that can be hoped for from continued practice and exercise, every day greater and more important advances, as is seen to be followed in all the wonderful and subtle inventions found by human ingenuity, both in the arts and in the sciences.

[5/p423] Galileo's General Report (1612)

It is well known to everyone who understands astronomical and geographical things, that, up to this age, no other way has been found of knowing the differences of longitudes of very distant places, both at sea and on land, except by the difference of hours, which are numbered in different regions at the same time, that some eclipse of the Moon or Sun is seen, but much better with those of the Moon, since they are more frequent and apparent to all at the same time.

By this single means all nautical and geographical maps and charts have hitherto been described; which, however, are found to be scattered with great errors, and especially those of the West Indies and all other very distant regions: and this, in my opinion, is due not only to the brevity of time in which such provinces were begun to be explored, and to their remoteness, which does not permit a continuous and frequent correspondence of observations, but also to the rarity of lunar eclipses; of which scarcely one or two occur every year, and which are often prevented from being observed by the cloudy air, and still more by the difficulty that different and distant observers have in noticing the same moment in the duration of an eclipse, which will be 1, 2, 3, and even sometimes 4 hours or more.

This use of eclipses, which, for the reasons given, is very time-consuming and uncertain, even for exact geographical descriptions, then remains altogether null in the same act of sailing over vast and remote seas; then that not once a year, but almost every day, it would be necessary to know punctually in what longitude the ship is, in order to be certain, by means of it and its latitude, of the punctual place it occupies over the globe.

This alone was lacking in the complete perfection of so great and useful the art [of localization], and this is what I have found, and I offer it to His Majesty: to whom I will give also some general information, so that the more easily may be lent an ear to what I am about to demonstrate and particularly explain in due time, when that remains served to accept and like my performance.

I employ in this investigation celestial observations, but of stars no longer observed, nor seen by others before me, which have very rapid motions of their own, the periods of which I have by long watches and labors exquisitely found and calculated. These stars have between them conjunctions, separations, eclipses, and other accidents, which by an infinite interval exceed in the present matter the usefulness of the eclipses of the Moon. And while the eclipses of the Moon are so rare that, grouped together, we have not one per year that we can use, of these we have more than a thousand per year that are very useful, so that no night passes that we do not have 2, 3, and even sometimes 4 and more.

As for exquisiteness, then, the phenomena are all so momentary and quick that, whether conjunctions, separations, occultations, apparitions, or eclipses, each one dispatches itself in a moment of time, so that in their measurement one can never err by half a minute of an hour; and in sum, they are so exact that no person of mediocre intelligence will not be able to describe in this way all the places of the world on maps and charts, with maximum error of 4 miles, even in the most remote regions. In addition, again by means of ephemerides calculated by me hour by hour, in which are contained for long times to come the times of said conjunctions, separations, eclipses, you will come in the same navigation, at whatever time of the night you like, in certainty of the true longitude, and in consequence, of the true place where the ship is; and this for ten months of each year, let it happen that for two months at most there remain such new invisible stars that it is at that time that the Sun is near them.

I will show to His Majesty and to those who He will designate, the appointed stars; I will show their movements, the continual mutations of aspects, that is, conjunctions, separations, eclipses, and other accidents, evening by evening, as far as it pleases Him, predicted and drawn by me long in advance, so that everyone may remain assured of the certainty of my predictions and of the correctness of my tables and calculations; I will teach not only the use, but the composition of these tables, and the way of adjusting them in all the centuries to come; I will explain the application of these celestial observations to the exact and accurate description of all the kingdoms of His Majesty, and of all the continents, seas, and islands of the world, and finally the way of making use of such inventions of mine even in the same navigation, so that others may at all times be certain of the place where they are to be found: an invention proportioned only to the greatness of the Crown of Spain, which alone with its kingdoms surrounds the whole Earth.

[11/785] Orso d'Elci to Belisario Vinta in Florence (Madrid, October 16, 1612)

[…] Galilei's invention of being able to sail from East to West has already been proposed here by another Spanish mathematician, who offered to test it. Until this fact has been clarified by the mathematician to the ministers, they cannot consider new propositions. […]

[12/1197] Galileo to Curzio Picchena, Prime Minister of Tuscany, in Florence (Rome, April 23, 1616)

[…] Here is the Rector of Villa Hermosa, secretary of the Most Excellent Count of Lemos, by whose order he came to see me; and among other things we discussed my invention of longitude. In six days, he will return to Naples, and from there to Spain.

But I shall do nothing without the renewed consent of His Highness, nor without the advice and favor of your Lordship, as I shall better say in due time by word of mouth, since, as I have said, I cannot write at length without harm and pain.

I reverently kiss your hands, together with the Reverend Priest Scarperia, and beseech you to continue to favor me with some of your commandments.

[12/1251] Galileo to Curzio Picchena in Florence (Pisa, March 22, 1617)

[…] While I was in Livorno, there were no ships outside the wharf, so I could only test the eyeglass on a small shuttle inside the wharf, where the water was relatively calm despite the strong wind. The minimal motion did not hinder the use of the eyeglass, even without an instrument to counteract the movement, which gives me confidence that I can overcome all challenges with the help of two devices I have devised.

I have already constructed a prototype of one of these devices [for making observations from a boat], and it is currently in the arsenal; I plan to test it as soon as possible. This particular device is not the one I believe will ultimately be used on ships to measure longitude, but I built it because I think it will be highly useful for His Serene Highness's galleys, allowing them to spot and identify ships at sea while underway. Signor Cavalier Barbavara and Maestro Lorio, with whom I have discussed this matter extensively, share my hope. They have explained the great advantage it would provide to our galleys if they could occasionally use the eyeglass, so I have worked to simplify its use to match the abilities of the seamen. I am nearly certain I can achieve this, provided those who will use it are willing to undergo eight to ten days of training that I will provide. For this, it is essential that those in authority command them, as this is a service of immense importance.

When the Lord Admiral arrives, it might be beneficial for Your Most Illustrious Lordship to encourage Their Most Serene Highnesses to express interest in testing what can be accomplished while I am here. This would allow me the opportunity to conduct tests on a galley, with Signor Cavalier Barbavara readily volunteering to accompany me to Livorno and assist in any way possible.

I wanted to inform Your Most Illustrious Lordship of this, leaving the decision to your wisdom. I humbly bow to Their Most Serene Highnesses, wishing them, as well as Your Most Illustrious Lordship and your beloved Lady daughter, a blessed Easter.

I remain, as always, your most devoted servant.

[12/1260] Galileo to Orso d'Elci in Madrid (Florence, June 30, 1617)

While I was thinking how to overcome the obstacles delaying the completion of my discovery regarding navigation by longitude, I stumbled upon another invention of immense utility for navigating galleys at sea. I hope to leverage this invention to expedite the resolution of the former matter with His Majesty.

I will briefly describe this new invention to Your Excellency, as well as the approach I propose for securing His Majesty's support.

Three months ago, while in Livorno with the Admiral and some galley captains, we discussed the significant advantage it would provide to our galleys if they could use a scope while sailing to spot and identify enemy vessels from a distance, assessing their type, number, and strength long before they could detect us. This would allow us to make informed decisions about whether to engage or evade, significantly enhancing our security.

However, they expressed concern that such a scope would be rendered ineffective by the constant motion of the galleys, especially at the mast's top, where the movement would make it impossible to keep the sight fixed on an object for even a brief moment.

I took this challenge upon myself and eventually devised a new type of scope that allows one to locate objects with the same speed as the naked eye and follow them without losing sight. This innovation magnifies the view more than tenfold, so that what is naturally seen at a distance of one mile can be viewed clearly at a distance of one hundred miles. Moreover, it can be used with both eyes simultaneously, providing great ease and enjoyment to the user.

This invention has been so highly valued by these Officers that we decided to keep it secret to prevent it from falling into enemy hands.—and they have assigned two noble knights to use this instrument at the mast's top, where usually only lowly crew members are stationed, who cannot be trusted with such a crucial task. The device is designed to be kept hidden, so only the user can understand its operation.

Additionally, this instrument offers another significant benefit appreciated by the Lords skilled in maritime affairs: it allows for the immediate determination of the distance between us and discovered vessels without any effort or loss of time.

Regarding the approach in advancing my other discovery concerning longitude with His Majesty, I have considered the following. Your Excellency has informed me that His Majesty, having previously spent large sums in advance on the mere promises of others who offered similar inventions, only to be disappointed, has resolved not to make further payments without assurance of success. To this, I have little to add, except that it would be unfitting for my reputation and resources to undertake a long and costly journey to present something of great utility and desire to a distinguished prince, only to face the potential obstacles and frustrations often encountered due to envy or malice from others who influence the great lords. Nevertheless, to ensure that His Majesty's time is not wasted and to allow me to proceed with less inconvenience and greater respect, I propose the following. I have obtained permission from the Grand Duke, my Lord, to offer His Majesty this latest invention, already perfected and secured for the safety of His Majesty's galleys. I request a payment of 1500 ducats to cover the costs of my journey, stay in Spain, and return, as well as the expenses for the necessary instruments and assistance required to complete the longitude project. I will provide His Majesty with every guarantee of my commitment, backed by the Grand Duke's word. Thus, as Your Excellency can see, the risk remains entirely mine, and His Majesty is only asked to reward a highly useful invention. Even this reward is modest compared to the utility derived from the invention. However, my desire to complete the extremely important longitude discovery compels me to act swiftly, as time is of the essence given my age and physical condition.

Now, regarding the potential difficulties Your Excellency mentioned, which may arise from those who will judge and implement my discovery: as for the challenges inherent to the invention itself, rest assured that any sudden concerns others might have encountered over years of handling this matter have already been resolved by me through continuous study and practice. This is not an invention that accidentally fell into the hands of someone unqualified, but a discovery by someone who has devoted his entire life to these studies. Therefore, it is improbable that I would make the same errors often committed by those who, lacking a true understanding of the relevant sciences, attempt to achieve conclusions that are impossible by nature and are easily recognized as such by experts.

I assert that I have overcome all the inherent difficulties of this discovery, which were numerous and far greater than those that might superficially arise. Your Excellency mentioned concerns about my claim of using stars invisible to the naked eye, suggesting that it seems impossible or impractical to locate them in the sky. I assure you that these invisible stars can be found as easily as the brightest stars, the Moon, or the Sun, because they are always near one of the largest stars in the sky. Finding one automatically leads to finding the others. I am confident that any other objections raised would be similarly addressed if they were presented to me through Your Excellency. However, it is unreasonable to expect that this operation be so simple that even the most foolish person could understand and practice it without study or training. It would be absurd to demand that something which has perplexed countless brilliant minds should be reduced to a triviality. Many arts, far less necessary than navigation, such as painting, sculpture, and music, require years of study, yet attract many practitioners. Why, then, should it be expected that the essential art of navigation should be mastered with such ease? My discovery, which enables the daily determination of longitude on sea and land, required me to find a method to significantly enhance vision—by thirty or forty times beyond natural limits. I achieved this, and it is a remarkable accomplishment. Yet even this would have been futile if nature had not placed certain wandering stars in the heavens, invisible to all before me, whose frequent changes in position could serve our needs. I discovered these stars, a noble finding, akin to discovering a new small world within our great world. However, even this was not enough without determining their exact movements and periods, which I achieved through five years of continuous observation, to the great detriment of my health and risk to my life. But all of this was in vain unless I applied this knowledge to navigation, addressing the practical challenges of implementing it. I accomplished this as well. However, expecting this operation, based on such advanced principles, to be understood by the simplest minds is unrealistic. The effort required to acquire some knowledge of astronomy and calculations to create tables year after year is something I must undertake, and not the sailors, who should be provided with the tables ready-made. And if I am not available, even while still alive, I will give the rules for calculating these tables to other astronomers. These rules and theories will never be lost, just as the others concerning celestial movements have not been lost and will not be lost, even though Ptolemy, Alfonso, and other inventors and scholars are no longer alive.

And this is regarding the difficulties that might arise in the matter itself, which I truly consider insignificant, while, on the other hand, I highly regard those difficulties which, although not at all related to the essence and reality of this matter, might be raised by someone who, either out of lack of understanding, envy, or some personal interest, might try to obstruct and disturb it. If such a person had great authority and influence with His Majesty and these principal Lords, and they completely deferred to his judgment and report, that could be problematic. However, I would not be afraid of this either, as long as His Majesty and the Great Lords themselves would be willing to dedicate some time to understanding this knowledge; because, absolutely, in a very short time, with discourse, reasoning, and sensible experience, I could make them fully capable and satisfied. But if it is unavoidable to be subjected to the judgments of others (something that I would not only not avoid, but would actually seek, provided that I deal with intelligent and sincere-minded people), I certainly request that any contradiction and opposition that others might want to raise against me be given to me in writing, so that I could use it on every occasion for my justification before the world. In this way, not only from the outcome, as is usually the case, but also from my proposals and the opposition of others, the world could better understand and make a more just judgment of my work.

Finally, as for the reward His Majesty plans to grant for this invention, the 2000 ducats of annual pension mentioned by Your Excellency is much less than what I had understood in Rome at the house of the Most Illustrious Cardinal Borgia, which was 6000 ducats with a cross of St. Iago. I ask Your Excellency to verify this information, and if true, to establish it accordingly. If not, I trust that Your Excellency will negotiate the best possible terms for me, ensuring the reward is fitting to my reputation. The minimum acceptable reward should be 4000 ducats annually for my lifetime, to be reduced to 2000 for my heirs after my death, along with the honor of Knight of St. Iago, if indeed it was the intention of His Majesty and his predecessors to honor the finder of this method with such a title.

Postscript: As I have written to Your Excellency before, I revived this matter in Rome at the house of the Most Illustrious Cardinal Borgia, in discussions with the Rector of Villa Hermosa, Secretary of the Excellent Count of Lemos. The Cardinal, expressing a desire to support this endeavor, advised me through a Roman Knight, a close friend of mine, to make use of his favor. However, I decided, and Signor Picchena agreed, not to pursue this favor but to rely on Your Excellency's support. The Cardinal may have written to the principal Lords of the court or even His Majesty himself, and his report and information about my situation may lend additional credibility to my project. I hope this serves as a useful notice to Your Excellency.

[12/1277] Orso d'Elci to the Duke of Lerma, Prime Minister of Spain (Madrid, September 11, 1617)

Most Excellent Sir,

Galileo Galilei, a native of Florence and a renowned mathematician throughout Italy, claims to have discovered, through his extensive studies and speculations, a method for calculating longitude, which would greatly benefit navigation to the Indies—a much-needed advancement that has yet to be achieved.

He assures that he can quickly and easily instruct pilots and sailors in the use of an instrument he has invented for this purpose and is willing to accept the same reward that His Majesty has offered to others for similar endeavors. Galileo requests only to be granted permission and guidance on how to travel to Spain, and to be provided with the means to support himself and the necessary officers, as well as to cover the expenses for the instruments required by this invention. For all these needs, including his return to Italy with his officers, he asks for a total of three thousand shields.

To ensure that this support and reward are not given in vain, Galileo offers to present, upon his arrival before His Majesty, another invention designed for the galleys and fleets of the Mediterranean. This invention consists of specially crafted glasses that, despite the motion and vibration of the galley, allow one to see enemy ships at ten times the distance of ordinary sight, enabling timely prevention of attacks or evasive maneuvers as needed. This instrument and these glasses have already been tested on the galleys of my Lord the Grand Duke, with successful results, while maintaining their secrecy.

Your Excellency may wish to consider whether His Majesty would be interested in testing such a significant invention, which could greatly enhance the security of Ocean navigation. The Grand Duke is prepared to grant permission for Galileo to travel to Spain and proceed to Lisbon or Seville, as His Majesty may direct.

[12/1283] The Duke of Lerma to the President of the Council of the Indies in Seville (Madrid, November 6, 1617) [Original in Spanish]

His Majesty has reviewed the document from the Ambassador of Florence that was attached to my letter. He has expressed his intention to bring a renowned mathematician from that state, who has discovered a method to calculate longitude, which would greatly benefit navigation to the Indies and enhance the security of voyages through the Ocean. His Majesty has instructed me to forward this matter to your Lordship for examination by the Council of the Indies and to seek your opinion on the matter as deemed appropriate.

May God protect your Lordship.

[A4/p79] Judgment of João Baptista Lavanha on the Galilean Proposal of Longitude (Spain, April 1618) [Original in Spanish]

The mathematician Galileo Galilei claims to have observed four stars near the planet Jupiter using a long-range telescope, and he asserts that these stars orbit the planet regularly. He proposes to use their movements to determine longitude, which navigators refer to as "height from east to west."

Galileo first introduced this discovery during his mathematics lectures in Padua and later published a book titled *Sidereus Nuncius*. However, six years ago, Francisco Sitio, a Florentine, wrote and published a book that thoroughly refutes Galileo's claims using astronomy, perspective, and philosophy.

Even if Galileo's observations are true, his method for determining longitude may still fail for two main reasons. First, there is uncertainty regarding the consistent motion of the four stars he intends to use. Second, there are significant challenges in how navigators would be able to make the necessary observations to determine longitude.

Regarding the first reason, the observations Galileo has made over eight or nine years are insufficient to verify and establish the movement of these stars as a reliable rule. It is possible that while their movements appear regular now, they may not remain so in the future. These stars could potentially disappear from the sky, as has occurred in similar cases before. Notable astronomers have achieved knowledge about planetary and stellar movements by comparing their observations with those of other experts, as Ptolemy did with Hipparchus, King Alfonso with Ptolemy, Copernicus with both, and Tycho Brahe with all of them.

As for the second reason, even if we assume that these four stars exist and their motion is consistently regular (which remains uncertain), the method Galileo proposes is impractical for navigators. The difficulty lies in the use of the tables that

Galileo would need to compile. These tables would have to indicate the longitudinal difference between the observation point and the reference meridian—whether it be Florence, Seville, or Lisbon—based on the position and aspect of the stars relative to Jupiter. Considering the general lack of education among navigators, many of whom cannot read or are unfamiliar with using tables like those for the Sun's declination, this method seems overly complex.

The impossibility arises from the practical use of the telescope itself. Even on land, it is challenging to observe the Moon, which appears much larger than the brightest stars. How then could one expect to accurately observe the four stars (which are invisible without a telescope) and their aspects relative to Jupiter while at sea, with the ship constantly moving?

In conclusion, I find Galileo's proposal unfeasible and ineffective.

Whether he shares it with the Dutch seems inconsequential, as their navigators could likely figure out how to use it, just as ours could if it were practical. This invention does not seem to rely on any secret of nature that cannot be understood by others.

The use of the telescope to sight distant ships is indeed useful, but it can be employed without relying on Galileo's method.

Your Majesty will ultimately decide what is most beneficial for the Kingdom in this matter.

[A4/p76] Resolution of the Spanish Council of State (April 28, 1618) [Original in Spanish]

The Council reviewed the letter Your Majesty sent through the Duke of Lerma, which included a memo from the Ambassador of Tuscany. This memo detailed an offer from Galileo Galilei, the mathematician to the Grand Duke and principal lecturer at the University of Pisa. Galileo proposes a method for calculating longitude that would greatly aid navigation in the Ocean, and he is willing to accept the reward Your Majesty has established for others who have made similar proposals. However, since Galileo would need to come here to demonstrate this method and teach it to sailors, he requests funds to cover his own expenses and those of the officers accompanying him. He asks for three thousand scudi to be provided as a guarantee for this purpose, to be used in Seville or Lisbon as arranged.

Additionally, Galileo offers another invention for Mediterranean galleys, which would allow them to sight enemy ships at ten times the distance of ordinary vision.

We ask Your Majesty to decide the most appropriate response to the mathematician's proposal.

Before forming an opinion on this matter, the Council sought the expertise of João Baptista Lavanha, who has experience in these areas. Lavanha reviewed the matter and submitted the attached report, in which he points out flaws in Galileo's proposal.

After considering this report, the Council asked Lavanha if he could replicate the telescope Galileo proposed, which he believed to be a useful tool. Lavanha responded that lenses would need to be imported from Venice, or Galileo would need to send the telescopes directly, as the type of glass needed is not available here. The Council concurred with Lavanha's assessment.

It seems to us appropriate to respond to the Ambassador of Florence with kind words, expressing appreciation for his good intentions and dedication to serving Your Majesty, and assuring him that Your Majesty will take the appropriate action.

[A4/p82] Appeal of Galileo (1619) [Original in Spanish, Translated from Italian by the Tuscan Embassy in Madrid]

Galileo Galilei, First Mathematician and Philosopher to the Grand Duke of Tuscany, wishes to remind Your Majesty that some years ago, he offered to come to Lisbon or Seville to teach the method of navigating to the Indies using the longitude he claims to have discovered. He has yet to receive a response from Your Majesty regarding the assistance he requested to support his travel and stay.

Since Galileo is eager to ensure that his invention is not lost and to demonstrate its effectiveness, he humbly requests that Your Majesty order that experienced navigators in Genoa or Naples provide him with the opportunity and means to conduct a test of his method. Galileo is willing to travel to these locations at his own expense, and the test would be carried out and verified by these experts.

Furthermore, he respectfully asks that Your Majesty grant him the honor promised to those who teach such navigation, which he will gratefully accept.

[A4/p83] Further Resolution of the Spanish Council of State (January 25, 1620) [Original in Spanish]

In 1618, the Ambassador of Florence proposed that Galileo Galilei offers a method for calculating longitude and ensuring safe navigation in the Ocean, requesting only the same reward given to others who had made similar proposals.

To assess the validity of this offer, the Council consulted João Baptista Lavanha, who, after consideration, concluded that the proposal lacked a solid foundation. However, he noted that the telescope suggested for implementing the method was valuable, recommending that lenses be procured from Venice due to the lack of suitable glass locally. The Council concurred with Lavanha's assessment and suggested that the ambassador be thanked for his diligence and intentions, noting that His Majesty was pleased to approve the proposal, as detailed in the attached consultation.

The Council recommends proceeding with the decision to allow Galileo to conduct his experiment, advising that arrangements be made in Genoa or Naples, where he will travel at his own expense. Furthermore, they request that Galileo be granted the honor previously established for those who teach navigation techniques.

[13/1439] Galileo to Giuliano de' Medici, Ambassador in Madrid (Florence, September 7, 1620)

For many years, I have offered to His Catholic Majesty my invention for determining longitude at any time and place—a matter of great importance for the accurate description of all the provinces of the world, for nautical charts, and for navigation itself, something sought after in every century but not yet discovered by anyone. My offer has encountered many difficulties, which have long delayed it from being given attention and being embraced according to the merit of its greatness. The main reason for this (as far as I have understood) is that in the past, many inventions were proposed which, once accepted and tested in practice, proved to be useless and of no value. As a result, His Majesty, having been defrauded many times, found that he had spent large sums of money to no avail. For this reason, it was decided to proceed much more cautiously and carefully in the future. However, my confidence in my discovery has led me to resolve to freely describe to His Majesty the principal foundation of this discovery, certain that His Majesty will appreciate my gift. The description follows.

Finding longitude is nothing more than, being in any part of the sea or land, knowing how far we are, to the west or east, from a meridian arbitrarily chosen by us as the point of reference from which longitude is measured.

Until this age, all ancient and modern geographers have come to this knowledge only through lunar eclipses, which are observed at different hours of the night from different parts of the Earth. For example, if the same eclipse that is seen in Seville ten hours after noon is seen in the Azores eight hours after their noon, it is clear that the Sun reached the meridian of the Azores two hours later than it did at the meridian of Seville, and consequently, these islands are thirty degrees more to the west. Now, if eclipses occurred every night, and their times of appearance in a specific place were calculated and tabulated, there is no doubt that sailors could know their longitude each night; but since eclipses are very rare, their use for navigation remains minimal and almost nonexistent.

But what has remained hidden until our age has fallen to me to discover: how in the sky every night, there are cosmic events observable worldwide, as suitable for determining longitude as lunar eclipses, and even more so. This is thanks to the four Medicean planets, which, in different orbits, continuously revolve around the star of Jupiter. By either converging two of them together, joining with Jupiter itself, separating from it, or eclipsing by falling into its shadow, or by emerging from that shadow, they give us, at various hours of each night, one, two, three, and sometimes even four or five significant points for the knowledge we seek. These are much more precise than lunar eclipses, as they are, in a way, momentary, so that the longitudes can be determined with an error of less than a mile.

These stars had been invisible and unobservable to everyone until now. With the excellent telescope that I invented and built, I have discovered them and observed them continuously for twelve years. Through long and laborious vigils, I have determined their movements and periods, and I have created tables with which I can calculate their conjunctions, eclipses, and the other aforementioned astronomical events at any future time, enabling me to accurately determine my longitude every night and in every part of the Earth and sea—every night, that is, when Jupiter is visible, which occurs throughout the year except for those days when it is under the rays of the Sun.

The endeavor is immense, and perhaps few surpass it in nobility, for it rests upon and is founded on three great wonders that I had to investigate. The first was finding an instrument that multiplies sight forty or fifty times beyond its natural capacity; the second was discovering a new world in the sky, with four new planets revolving around it; the third was determining the periods of revolution of all four, so that through them I can accurately calculate their astronomical events.

This is a brief account of my progress, well worthy of His Catholic Majesty, for whose greatness new parts of this lower world are discovered, and entire new worlds are unveiled in the heavens.

Appendix B: Documents Related to the Presentation of Galileo's Proposal to the States General of the Netherlands

[16/3339] Galileo to Admiral Laurens Reael in Amsterdam (Arcetri, August 15, 1636)

Having resolved to present my invention for determining longitude, a crucial and highly sought-after element for the perfection of nautical art, to the Most Illustrious and Most Powerful General States of the Belgian Provinces, I found myself lacking a person of great intelligence and experience in the field—someone of sincere character, well-regarded by those same Lords, who could both present and, if necessary, protect my invention.

The renowned reputation of Your Illustrious Lordship, which extends beyond even the extensive boundaries of these famed provinces, reached my ears, reinforced by such testimonials of your great virtue and kindness, that it gave me the courage to seek your help and support to introduce my work with the dignity appropriate to such distinguished and eminent powers.

The confidence that Your Illustrious Lordship extends to me, which could not be given due to my humble status, comes instead from the importance of the subject and proposal I am presenting. Your Illustrious Lordship is well aware of the significance of safely navigating the vast ocean, having skillfully commanded numerous fleets more than once. Thus, I am sending you the free and clear presentation of my invention for the Most Illustrious and Most Powerful Lords. I am sending it to you open, so that you may first review and consider it yourself, and, finding it neither vain nor unworthy to be presented to the most prudent Lords, present it on my behalf. If my own affection has misled me, let my good intentions alone be appreciated and the writing suppressed.

I wish to emphasize to Your Illustrious Lordship, as someone who understands this better than I, that all great and noble arts began from modest and simple principles. If the early inventors had not been succeeded by speculative minds, who with keen intellects understood that these modest beginnings contained the foundations

A. De Angelis, *Galileo and Satellite Navigation*, SpringerBriefs in History of Science and Technology, https://doi.org/10.1007/978-3-031-78799-7

of wonderful arts, those arts would, as the saying goes, have died in their infancy, and the world would have remained in a state of rough, uncultivated ignorance.

There are countless examples of this, as many as there are noble and industrious arts. If we consider the marvels of various musical instruments perfected over time, what difference is there between these and the first lyre of Mercury or Pan's syrinx? What shall we say of the art of weaving, which began with the simple act of braiding a mat? And now, in particular, your people can send men across all islands and continents to make the necessary observations, first to correct all geographical descriptions, and others to study the composition of ephemerides, and others to practice using the telescope.

I have given a brief account of this initial presentation and information. From this, the Most Illustrious Lords may decide, with the advice of knowledgeable scientists and astronomers, what actions to take in this matter. For the brief time that my life may last, I remain fully ready to provide anything that might be needed to perfect this noble undertaking. Meanwhile, I ask Your Illustrious Lordship to appreciate the confidence I have placed in your favor, even if it may not seem deserving. Where my merit falls short, let the significance of the proposed endeavor make up for it, and let your kindness value my offer and dedication.

With all due reverence, I kiss your hands and wish you every happiness and greater glory.

[16/3337] Galileo's Proposal to the States General of the United Provinces of the Netherlands (August 15, 1636)

To You, Most Illustrious and Most Powerful Lords, esteemed rulers and masters of the Ocean, fortune, indeed divine providence, has granted the opportunity to advance the art of navigation to its utmost perfection. As experts know well, despite the impressive achievements of your renowned nations, there remains one critical area still sought after: the ability to determine longitude as accurately as latitude. This capability is essential for precise location determination both at sea and on land across our vast globe.

For centuries, astronomers and other scholars have sought a method to determine longitude, with great honors promised to those who would succeed. Up until now, the only known method has been the ancient one involving lunar eclipses, used by geographers to create their charts. However, due to the rarity of such eclipses, this method remains largely impractical for mariners. It is impossible to determine the difference in longitude from events that occur on Earth, except uselessly between nearby locations, because neither smoke signals by day nor fires by night can be observed even at a distance of one degree. Therefore, one must resort to very high and celestial events that are visible across an entire hemisphere. The heavens were generous with such events in past ages, but for our present needs, they are quite scarce, helping us only with lunar eclipses. Not that the heavens themselves are not abundantly full of frequent and notable occurrences, more suitable and fitting to our needs than lunar or solar eclipses; but it pleased God to keep them hidden until our times, and then reveal them through the ingenuity of two men, one Dutch and the other Italian, Tuscan, and Florentine: the former, as the first inventor of the telescope or Dutch tube; and the latter, as the first discoverer and observer of the Medicean Stars, so named after the dynasty of his Prince and Lord. Now, to get to the point in brief words, I present to Your Most Illustrious and Powerful Lordships the whole story and summary of this matter.

Let it be known, then, that around Jupiter's body, four smaller stars revolve perpetually with different velocities in four circles of varying sizes; from the movements of these stars, we have, for each natural day, 4, 6, 8, and often more astronomical events, each of which is no less useful, indeed much more so, than as many lunar eclipses for the investigation of longitudes, given that their short duration leaves no room for error in counting the hours and their parts. The events are as follows: First, since the body of Jupiter is by nature no less dark than the Earth and shines only by the illumination of the Sun, it casts its shadow in the form of a cone on the side opposite the Sun, through which each of its four satellites passes as it moves along the upper part of its circle; and since these satellites, like four moons, are also devoid of light and shine only by the Sun's illumination, they are eclipsed when they enter the cone of Jupiter's shadow, and because of their small size, the immersion into darkness occurs in about one minute.

Likewise, some hours later, upon emerging from the shadow, they regain their brightness in a very short time: from this, it is clear that observers of such eclipses cannot differ from each other in the exact time of observation by more than a minute. In addition to the eclipses, there are also the moments when their bodies come into contact with that of Jupiter; where one can observe the exact moment when they appear to touch Jupiter's disk, and conversely, when they separate from that disk: such conjunctions and separations can be observed without an error of half a minute, thanks to the speed of their motion and the very short time between touching and not touching. Thirdly, the conjunctions and separations between these satellites themselves can be observed, as they move in opposite directions, bringing them to an exact conjunction, which occurs in less than a minute, so that its midpoint can be measured very accurately, without error even of a few seconds.

These are the frequent events of every night, in any part of the entire globe, and at all times of the year when Jupiter is visible and observable: from these events, when ephemerides are created by a skilled astronomer and calculated for a specific meridian, such as the meridian of Amsterdam, and when sailors have copies of these with them, by making timely observations and comparing them with the times noted in the ephemerides, they will be able to determine the distance of the meridian they are on from the prime meridian of Amsterdam, which is the sought-after longitude.

The great security and usefulness of being able to correct and amend all geographical and nautical charts on land, so that they do not differ from the truth by even half a degree or, I would almost say, a mile, is very clear and easy; because, without ephemerides or other calculations, it is enough for a person, in the place where they are, to observe the above-mentioned events for a few nights, noting the time of their appearance, which, when compared with the same observations made and noted, with their times, in Amsterdam or another place, will give the difference in meridians: so we are certain that this practice will be exercised in the future; and with it, all of geography will be restored to absolute accuracy, achieving in fewer years what has not been achieved in many centuries with the help of lunar eclipses. However, for the use of navigation, there are 4 particular requirements to be met. First, the precise theory of the movements of these Medicean circumjovian stars, so that skilled astronomers can calculate and distribute all the above-mentioned events into ephemerides. Secondly, telescopes of such perfection are needed, to make these stars clearly visible and observable. Thirdly, a way must be found to overcome the difficulty that some might believe is posed by the movement of the ship when using the telescope. Fourthly, an exquisite clock is needed to count the hours and their fractions, from noon or sunset.

As for the first point, I have calculated the periods of the movements of the four stars with such precision that the astronomical events, calculated many months in advance, correspond exactly; and (as those skilled in observations and celestial motion calculations know) the passage of time continually adds greater accuracy.

As for the second point, I have perfected the telescope to such an extent that Jupiter's satellites, though invisible not only to the naked eye but also to ordinary telescopes, appear no less large and bright than fixed stars of the second magnitude seen with the naked eye; indeed, they continue to be visible even at twilight when none of the fixed stars are still visible. I also believe that similar or even greater perfection in telescopes may be found in those regions where the first invention originated.

Regarding the third point, I have also considered a suitable remedy to place the observer in a position so well-prepared that they will not feel the movement of the ship. However, considering how many operations have been discovered through the progress of time, experience, and the ingenuity of human minds, I have no doubt that the practice of careful and patient men will develop this method for use at sea as well as on land, especially since our operation does not involve taking distances with quadrants or other such instruments between stars, but simply passing the sight to see if two of those satellites are aligned, if they are in contact with Jupiter's disk, or if they have exited or are about to enter the cone of shadow; of which events, having been forewarned by the ephemeris that they will occur on that night, by frequently repeating the observation, they will precisely catch the time and hour of the event.

Finally, concerning the fourth requirement, I have such a time measurer that if 4 or 6 of such instruments were made and left to run, we would find (to confirm their accuracy) that the times measured and displayed by them, not only from hour to hour but from day to day and month to month, would not differ from each other by even a second of a minute, so uniformly do they operate: truly admirable clocks for the observers of celestial motions and phenomena; and the construction of such instruments is exceedingly simple and straightforward, and much less prone to external alterations than any other instrument found for similar use.

I know very well, Most Illustrious and Powerful Lords, that in front of great Princes one should come with new inventions already established and ready for immediate use; however, I also know that your prudence will understand that, not being a mariner nor suitable for navigation, I could not present myself in front of you in any other way than this. Perhaps I could have appeared in person if the length of the journey, my advanced age of 73 years, and other impediments had not held me back. But what reassures me, beside your Lordships' kindness and magnanimity, is that I have claimed nothing else, except that your prudence and humanity will appreciate this small offspring of my intellect, which I freely offer as a gift, as well as offering whatever else may be needed for the full completion of this enterprise. And here, to conclude, I would like to add this: that your Most Illustrious and Powerful Lordships, being truly mightier than all other powers in the world to begin and perfect such a desired and sought-after undertaking, should not hesitate to apply your thoughts and hands to it; and be assured that now, or at some other time, this invention will be put into use, which can be called admirable, as it depends on celestial and divine matters, placed up there by God solely to benefit humankind. The beginnings of all great endeavors come with difficulties, which the patient industry of men overcomes with time, as anyone can clearly understand by considering the many arts, the beginnings of which we know were very weak, and now they are seen to perform feats that astonish even the most elevated minds. I could name countless arts, but let this one alone suffice: navigation, which your own Dutch have brought to such marvelous perfection; and if this one remaining skill, that of finding longitude, which seems reserved for them, is added to their other industrious operations as their ultimate and greatest achievement, they will have set the limit and goal of glory, beyond which no other nation can hope to pass.

And I humbly bow.

[17/1808] Galileo to Laurens Reael in Amsterdam (Arcetri, June 6, 1637)

Along with the highly esteemed and gracious letter from Your Most Illustrious Lordship, I also received a correspondence from the very distinguished and learned Martinus Hortensius. Both letters were sent to me by my dearest and most trusted friend, the exceptionally distinguished Signor Elia Diodati from Paris. Unfortunately, I was unable to read them due to a severe affliction affecting my right eye, which has rendered me nearly blind and forced me to rely on others to read the letters for me. This condition, caused by extensive writing over the past three months, also impedes my ability to write. Therefore, to better serve both Your Most Illustrious Lordship and Mr. Martinus Hortensius, I have decided to respond to you jointly, as the inquiries in your letters are similar.

Your Most Illustrious Lordship has informed me that my proposal has been presented to the Most Illustrious and Powerful Orders of the United Provinces and has been received with gratitude. Additionally, you mentioned that a copy of the resolution would be sent to me via Mr. Hortensius. However, I have not yet received this copy, nor the authentication from Mr. Cornelius Musch, whom I understand is the Chancellor of these Most Powerful Lords. Despite this, I am eager to address the questions and concerns raised about my method for determining longitudes at sea and on land.

The main doubt raised by Your Most Illustrious Lordship, as conveyed to me by Mr. Hortensius, concerns the possibility of using the telescope on a ship, which due to the fluctuations of the waves might not allow proper observations of the satellites of Jupiter. The second difficulty, also mentioned by Mr. Hortensius, is the lack of telescopes of sufficient perfection to clearly distinguish the small stars accompanying the planet Jupiter. Mr. Hortensius also requests tables and a method to accurately calculate, from time to time, the movements and consequently the aspects of these small stars. Additionally, he inquires about the construction of the clock I proposed, of such exquisite precision that it can measure even the smallest fractions of time without any error, in all places and all seasons of the year.

Regarding the first difficulty, there is no doubt that it is the greatest concern, for which I believe I have found a remedy during the moderate movements of the ship, and this should suffice, considering that in severe agitation and storms, which often obscure even the sight of the Sun, let alone other stars, all other observations cease, as do all nautical duties. Therefore, I believe that during moderate disturbances, the condition of the observer can be reduced to a calmness similar to the tranquility and stillness of the sea. To achieve this benefit, I have thought of placing the observer in a specially prepared location on the ship, where not only the motions from bow to stern but also the lateral ones from side to side are not felt at all. And my idea is based on this principle: if the ship were always in perfectly calm water, with no waves, there is no doubt that using the telescope would be just as easy as on land. Now, I want to place the observer in a small boat within the large ship, with this small boat containing a quantity of water as required, which I will explain shortly. Here, it is evident that the water in the small vessel, even if the large ship tilts or rocks to the right, left, forward, or backward, will always remain level without ever rising or falling in any part, but will always remain parallel to the horizon. So, if we placed another smaller vessel floating in the water within this small boat, it would find itself in perfectly calm waters, and consequently, it would remain without rocking. And this second small vessel is where the observer should be placed. Therefore, I want the first vessel, which must contain the water, to be like a large basin in the shape of a half-spherical orb, and that the smaller vessel should be similar in shape, only smaller, so that between its convex surface and the concave surface of the containing vessel, there is no more space than the thickness of a thumb. This way, a very small amount of water will suffice to support the inner vessel, no less than if it were set afloat in the wide ocean, as I demonstrate in my treatise on floating bodies [referring to *Discourse on Floating Bodies* published in Florence in 1612; see the third volume of the *Works* (Note by the Editor)], which indeed seems marvelous and incredible at first glance. The size of these vessels should be such that the inner and smaller one can support the weight of the person making the observations, along with the seat and other devices necessary for positioning the telescope, without sinking. And to ensure that the inner vessel is always separated from the surface of the containing vessel without ever touching it, so that it cannot be moved by the agitation of the ship, I want some springs to be fixed on the internal concave surface of the containing vessel, or on the convex surface of the contained one, in a number of eight or ten, to prevent the two vessels from coming into contact, but without preventing the inner one from moving with the tilting and rocking of the containing vessel. And if instead of water, we wanted to use oil, it would serve just as well, if not better, and the quantity needed would not be large; two or at most three barrels would suffice. Your Most Illustrious Lordship and Mr. Hortensius could conduct a small experiment with two small copper basins, placing a quantity of sand in the smaller one, as long as it floats on the water, and fixing a rod upright in the sand, then moving the external basin by tilting it now to this side, now to that; they will see that the rod remains in the same position without tilting, especially if the tilting of the containing basin is done slowly and with a noticeable interval of time between one and the other, as is finally the case with large ships. But Your Most Illustrious Lordship can rest assured that when one begins to study the practice of such operations, there will be no shortage of skilled individuals who will, over time, become accustomed to performing these operations without other artificial preparations. I once made, for use on our galleys, a certain cap in the form of a helmet which, when worn by the observer and having a telescope fixed to it in such a way that it always looked at the same point that the other free eye was directed towards, without doing anything else, the object viewed with the naked eye would always align with the telescope. A similar device could be constructed, which would not only be worn on the head but also fixed on the shoulders and torso of the observer, with a telescope of the necessary size to clearly discern the small Jovian stars, and it would be so well-adjusted to one of the eyes that by simply directing the sight to the body of Jupiter, the other eye would align it with the telescope, and consequently see the stars close to it.

As for the second point, which concerns finding telescopes more powerful than those made there, I believe I have already written that the one I have used has such a power that it first shows the disk of Jupiter not rough, but very well-defined, no less than the naked eye can see the limb of the Moon, and similarly it shows the satellites of Jupiter very clearly, with a size such that fixed stars of the second magnitude are not seen larger or more distinct with the naked eye. Moreover, by following Jupiter's movement with the telescope, these satellites can be seen in the evening before, and in the morning after, the appearance or disappearance of fixed stars. And Jupiter itself, when followed with the same telescope, can be seen all day long, as can Venus and the other planets, and a good part of the fixed stars. Here, Your Most Illustrious Lordship and Mr. Hortensius can judge what an immense benefit this marvelous instrument brings to astronomical sciences. I will not fail to send the lenses to Your Most Illustrious Lordship, and perhaps they will arrive with this letter, if my craftsman, who makes them, has the opportunity to make one for me. And I say this because the Most Serene Grand Duke, my Lord, being enamored with such instruments, keeps this man of mine constantly at his side, taking him with him wherever His Highness goes, whether in towns or villas. So there should be no doubt about the construction and success of these instruments.

Now I come to the second artifice to immensely enhance the accuracy of astronomical observations. I speak of my time measurer, whose precision is such that it will not only give us the exact amount of hours, minutes, and seconds, but even third fractions of time if we were able to count them; and the accuracy is such that if two, four, or six of these instruments were made, they would operate so precisely together that one would not differ from another, not only in an hour, but in a day, or even in a month, by even a single pulse; and the foundation of such construction I derive from a marvelous proposition, which I demonstrate in my book on motion that is currently being printed by the Elzeviers in Leiden, and the proposition is as follows: If in a circle erected on the horizon, the perpendicular drawn from the point of contact (which in consequence will be the diameter of the circle), and from the point of contact, or the top end of the diameter, as many chords as one wishes are drawn, upon which it is imagined that moving bodies descend as if on inclined planes, the times of their passages over such chords and over the perpendicular diameter itself will all be equal: so that if, for example, chords are drawn from the lowest contact point to the circumference, covering subtensions of 1, 4, 10, 30, 50, 100, or 160 degrees, the moving body will descend over all of these in equal times, and also over the entire perpendicular diameter. And this also happens in the sections of the circumferences of the two lower quadrants, where, as if they were channels down which a heavy globe descends, it will traverse the entire circumference of the entire quadrant in as much time as if it had started moving 60, 40, 20, 10, 4, 2, or just one degree away from the lowest point of contact. Truly a phenomenon full of wonder, which everyone can verify by suspending a small lead ball or other heavy material from a thread tied high above, and then moving it away from its vertical position until it is elevated by a quarter; letting it go, it will be seen to go back and forth, making many oscillations, large at first, then continuously diminishing, until it no longer moves more than a single degree to either side of its vertical position; and as it continues to move along the same circumference, one will see the large, medium, small, and very small vibrations all occur in equal times. And for a more certain experience, suspend two similar small balls from two threads of equal length, and if one is moved away by a very large arc of eighty or more degrees from the perpendicular, and the other by only two or three degrees, and then let go, one person counting the vibrations of one pendulum, and another counting the vibrations of the other pendulum, it will be found that when one counts a hundred of the large oscillations, the other will also have counted exactly a hundred of the very small ones.

From this very true and stable principle, I derive the construction of my timekeeper, not using a weight suspended by a string, but a pendulum made of solid and heavy material, such as brass or copper. I make this pendulum in the form of a sector of a circle with an arc of twelve or fifteen degrees, and with a radius of two or three palms; and the larger it is, the less tedious it will be to operate. This sector is thicker at the middle of the radius, tapering towards the outer edges, where it ends in a very sharp line to minimize air resistance, which alone would otherwise slow it down. It is perforated in the center, through which passes a small rod shaped like those on which steelyards rotate. This rod ends in a point at the bottom and rests on two bronze supports, so as to reduce wear from the prolonged motion of the sector.

When this sector is moved several degrees from the perpendicular position (when it is well-balanced), it will oscillate back and forth a great number of times before coming to a stop, and to continue the oscillations as needed, the operator will need to give it a strong impulse at the right time, restoring it to wide oscillations. And once the number of oscillations occurring in a natural day, as measured by the revolution of a fixed star, has been patiently counted, one will have the number of oscillations in an hour, a minute, and smaller units of time. It will also be possible, after this initial experiment with a pendulum of any length, to lengthen or shorten it so that each oscillation takes exactly one second; for the lengths of such pendulums maintain a squared proportion relative to the times, as for example: if a pendulum of four palms in length makes a thousand oscillations in a given time, when we want the length of another pendulum that in the same time would make double the number of oscillations, the length of the pendulum must be a quarter of the length of the other. And in sum, as can be seen by experience, the number of oscillations of pendulums of different lengths is proportional to the square root of those lengths.

To avoid the tedium of someone having to constantly count the oscillations, there is a very convenient device for this purpose; that is, by having a very small and thin needle extend outward from the middle of the sector's circumference, which, as it passes, strikes a fixed bristle with one of its ends. This bristle rests on the teeth of a wheel as light as paper, which is placed horizontally near the pendulum, and having teeth around its edge like those of a saw; that is, with one side positioned squarely on the plane of the wheel, and the other slanted obliquely. This arrangement will serve so that when the bristle hits the perpendicular side of the tooth, it will move it, but upon the pendulum's return, the same bristle, sliding along the oblique side of the tooth, will not move the wheel at all, but will fall onto the next tooth. And so, with each pass of the pendulum, the wheel will move by the space of one of its teeth, but the wheel will not move at all when the pendulum returns; thus, its motion will always be circular in the same direction. By marking the teeth with numbers, one will see, at one's discretion, the number of teeth that have passed, and consequently the number of oscillations and elapsed time intervals. Furthermore, around the center of this first wheel, another small wheel with a few teeth can be added, which will engage another larger toothed wheel, from whose motion we can learn the number of complete revolutions of the first wheel, distributing the number of teeth so that, for example, when the second wheel has made one revolution, the first has made 20, 30, 40, or as many as desired: but explaining this to Your Most Illustrious Lordships, who have exceedingly skilled and ingenious men in the construction of clocks and other marvelous machines, is unnecessary, because they themselves, with this new foundation of knowing that the pendulum, whether moving through large or small arcs, always makes its oscillations equally, will find subtler conclusions than I can imagine. And since the failure of clocks mainly consists in not having yet been able to construct what we call the clock's timepiece so precisely that it makes its oscillations equal; in this my simple pendulum, not subject to any alteration, lies the way to always maintain equal time measurements. Now Your Most Illustrious Lordship, together with Mr. Hortensius, understands the benefit for astronomical observations, for which it is not necessary to keep the clock running continuously, but it suffices, for counting the hours at noon or at sunset, to know the fractions of time until some eclipse, conjunction, or other aspect of celestial motions.

As for the tables of the movements of Jupiter's satellites, and the method I have used to calculate and create the ephemerides, I cannot fully satisfy these requests at the present time, because I am so greatly impeded by a flux in my right eye, which, to my greatest regret, prevents me from being able to write or even read a single word, and needing, by the grace of Mr. Hortensius, to review the current constitutions to establish the roots of these movements, in order to adjust their mean motions, and moreover to check the large number of observations I have made continuously over many years; unable to make any use of my sight, I must wait as long as my ill fortune pleases, which could perhaps be many days.

As for the part concerning the most learned and excellent Mr. Martinus Hortensius, that is, to be able to begin practicing my invention on land, to adjust maps and establish with utmost precision the longitudes of islands, ports, and other fixed locations; for this, there is no need for tables or other ephemerides, but it requires two observers, one fixed at the prime meridian, which I assume to be that of Amsterdam, and the other traveling from place to place making observations for three, four, or six nights of the conjunctions, separations, and other aspects, keeping an exact account of the time, which falls between noon and the occurrence of such aspects. These, when sent and compared with those occurring and observed in the same way, will give the difference in meridians, that is, the sought-after longitude. Therefore, before anything else, it will be necessary for the Most Illustrious and Mighty Lords of the States to commission an observatory in Amsterdam, equipped with the necessary instruments for continuous observations, and that a man skilled in astronomy, diligent, and patient, such as I have been for many years to discover what I have achieved through truly Herculean labors, be appointed to this office.

For such an office, I know there are capable men in those parts. However, from what I have been able to discern of Mr. Hortensius' abilities, I believe he would be not only highly suited for this service, but without equal, or at least without superior. Therefore, if this gentleman does not refuse to take on the enterprise, I will send him everything that remains to fully and freely reveal to the Most Illustrious and Mighty States all my inventions. And because what I am about to add is the most crucial point of this entire enterprise, I will not hesitate to repeat it, even though I have already written with great emphasis.

Your Most Illustrious Lordship should therefore bear with me as I repeat that not only have the beginnings of great enterprises and arts been small and required that the diligence and constant study of perceptive minds gradually overcome the initial apparent difficulties: but the same has occurred in the most minor and humble arts. I infer from this that, having not been able to present an art already established and perfected, since I have neither been a sailor nor an explorer of remote places, the Most Illustrious and Mighty States must defer to the judgment of knowledgeable persons, and if they wish to achieve the desired goal, they should command that this great enterprise be undertaken without interruption or delay due to the difficulties encountered at the beginning, because all of them will be overcome, as human ingenuity has overcome far greater obstacles.

I have chosen to present my invention to the Most Illustrious and Mighty States more than to any other absolute Prince, because when the Prince alone is not sufficient to comprehend the entire mechanism, as almost always happens, and must defer to the advice of others, who are often not very knowledgeable, that sentiment which rarely separates itself from the human mind, namely the unwillingness to see others exalted above oneself, causes the ill-advised Prince to scorn the offers; and the offerer, instead of receiving reward and thanks, encounters disturbance and disdain. But in a Republic, where decisions depend on the consultation of many, a small number, or even a single one of the Powerful, who is moderately knowledgeable about the proposed matters, can encourage the others to give their assent and participate in embracing the enterprises. This support I have hoped to receive from the favor and authority of Your Most Illustrious Lordship, and if it happens that by your advice the enterprise is taken up, I will feel great contentment, though my advanced age leaves me no hope of seeing my studies and labors yield and mature the fruit that I foresee will benefit humanity in these two grand and noble arts, Nautical Science and Astronomy. I have held Your Most Illustrious Lordship's attention too long; I beg you to forgive me and to share what I write with Mr. Hortensius and Mr. Blavius, chosen as the third of the Lords Commissioners, greeting them with reverent affection on my behalf, while I humbly bow to Your Most Illustrious Lordship and pray that God grants you the fullness of all happiness.

[17/3468] The States General of the Netherlands to Galileo in Arcetri (Amsterdam, April 25, 1637) [original in French]

It has been five months since Mr. Reael, former Governor General of the East Indies, presented to us your remarkable gift—the invention for determining longitude at all times, a long-sought solution that has eluded discovery for centuries. We have conveyed to Mr. Reael our deep appreciation for your contribution and our gratitude to you. We are also aware of the need to validate this invention through rigorous tests and evaluations conducted by our most learned mathematicians and experts in these regions. We eagerly wait for their findings.

As a token of our appreciation and goodwill, we are sending you gifts, including a gold necklace valued at approximately two hundred florins. Should your invention prove to be as effective as you have assured us, we will ensure that our gratitude is expressed in an even more generous manner, alongside the honor and acclaim that will be rightfully yours.

[18/3961] Galileo to Hug de Groot in Paris (Arcetri, January 15, 1640)

I am deeply sorry for the sudden death of Mr. Martinus Hortensius, which occurred after the deaths of the other three Commissioners. These misfortunes, added to my own injury, seem to be hindering and disrupting the continuation and progress of the project I have undertaken with the Most Illustrious and Most Powerful Lords States. The business, however, as far as it will be possible for me, will not remain impeded or delayed, since I have met a person most intelligent in these astronomical sciences and most apt to not only make up for the defect caused to me by blindness, but to carry it on with greater accuracy, being, in addition to his perfect intelligence, a man of nimble and shrewd wit, of complexity and freshness of age fit for all labor, of keenest sight, of an ardent desire to go forward, embracing the enterprise with great enthusiasm, and (what pleased me most) even willing to move to Holland if the Most Illustrious and Most Powerful Lords deem it necessary. He is prepared to send the ephemerides of Jupiter's satellites for the next six or eight months, calculated and drawn in advance. These ephemerides will provide exact predictions of the stars' positions night by night. By comparing these predictions with actual observations, experts will be able to confirm that we have achieved an accurate calculation of the Medicean Stars' movements, which is the foundation of this project.

I have given significant importance to this meeting to assure the Most Illustrious and Most Powerful Lords, and all experts in astronomy, that my proposal is and remains soundly founded. Therefore, I request that, through the means Your Very Illustrious Lordship finds appropriate, my thoughts be communicated to the Most Illustrious and Most Powerful Lords. This way, they might consider appointing new Commissioners to continue the project, and I, upon receiving notice from Your Lordship through other letters, would proceed with the remaining tasks. There is truly no reason to abandon such a significant project, given its immense utility, which cannot be achieved by any other means or invention, and which requires only time and minimal expense. The time needed could be greatly reduced if the project, currently managed through distant correspondence with considerable risk of loss, were handled directly and promptly by my colleague. My friend Diodati would not hesitate, for such an important undertaking, to move to your location or to Venice if the Most Illustrious and Most Powerful Lords chose to appoint, among others, the Illustrious Ambassador they have in Venice. There, proximity and the convenience of speaking directly with my friend would greatly facilitate the prompt and perfect completion of the project.

I will therefore wait for the response that Your Very Illustrious Lordship will have received from those parts.

I had previously written many months ago to the Most Illustrious and Most Powerful Lords, informing them that I accepted and appreciated the gift of the necklace with due thanks and reverence. However, I will not wear it until the project is completed. The necklace remains with the merchant who delivered it, and I left it in his care to avoid offending the magnanimity of those Lords.

I wanted Your Very Illustrious Lordship to be informed of these matters so that you can verify that my proposal is sincere and not made lightly.

[18/4021] Elia Diodati to Galileo in Arcetri (Paris, June 15, 1640)

I was deeply sorry, as can be confirmed by the Most Illustrious Signor Count Bardi, for the delay in writing to you due to waiting for letters from Holland. Since my last correspondence on February 17, I have not received any satisfactory response regarding your business from that location, despite Mr. Huygens's assurances. He had written to me urgently, as he is a principal figure in the State, being the first Counsellor and Secretary to the Prince of Orange, and holds considerable authority with him and the Lords States-General. You might have some insight into this from the translation of his letter that I am sending you.

Despite this hope, our wish has unfortunately not been fulfilled. Although, in line with your advice, I wrote to Mr. Borrel in Amsterdam over three months ago, I have yet to receive any reply from them. Consequently, I felt it was necessary to update you, having exhausted all possible efforts and knowing well the disappointment this new lack of response will cause you.

The Medicean astronomical tables you entrusted to me through Signor Conte Bardi have been reviewed and examined by mathematicians, all of whom approve of and commend the work. However, the material sent is incomplete, missing the final sections, and lacking tables from pages 12 to 25 regarding the Sun's motion. As a result, they cannot offer a fully informed evaluation.

Bibliography

Bedini, S. 1991. *The Pulse of Time. Galileo Galilei, the Determination of Longitude and the Pendulum Clock*. Florence: Olschki.

Camerota, M., and P. Ruffo, eds. 2019. *The Works of Galileo Galilei: National Edition, Appendix*. Vol. 4. Florence: Giunti.

De Angelis, A. 2021. *Discorsi e dimostrazioni matematiche intorno a due nuove scienze di Galileo Galilei per il lettore moderno*. Torino: Codice.

Dunn, R. 2012. Scoping Longitude: Optical Design for Navigation at Sea, in From Earth-Bound to Satellite. In *Telescopes, Skills and Networks*, ed. A.D. Morrison-Low, S. Dupré, and G. Strano. Leiden: Brill.

Favaro, A., ed. 1890–1909. *The Works of Galileo Galilei: National Edition*. Florence: Giunti.

Proverbio, E. 1984. *The Determination of Longitudes at Sea in the Seventeenth Century and the Contribution of Galileo*. Cagliari.

Van Helden, A. 1993. *Longitude and the Satellites of Jupiter, Proceedings of the Longitude Symposium*. Cambridge, MA: Harvard University.

Vanpaemel, G. 1989. *Science Disdained. Galileo and the Problem of Longitude*. In *Italian Scientists in the Low Countries*, ed. C. Maffioli and L.C. Palm. Amsterdam: Rodopi.

© The Author(s), under exclusive license to Springer Nature Switzerland AG 2024
A. De Angelis, *Galileo and Satellite Navigation*, SpringerBriefs in History of Science and Technology, https://doi.org/10.1007/978-3-031-78799-7